The European Union's Energy Security and Turkey's Role in the Southern Gas Corridor

Faik Tanrıkulu

The European Union's Energy Security and Turkey's Role in the Southern Gas Corridor

Interdependence on the Natural Gas Pipeline between Turkey and EU

Bibliographic Information published by the Deutsche Nationalbibliothek
The Deutsche Nationalbibliothek lists this publication in the Deutsche National-bibliografie; detailed bibliographic data is available in the internet at http://dnb.d-nb.de.

Library of Congress Cataloging-in-Publication Data
Names: Tanrıkulu, Faik, 1983- author.
Title: The European Union's energy security and Turkey's role in the Southern Gas Corridor : interdependence on the natural gas pipeline between Turkey and EU / Faik Tanrıkulu.
Description: New York : Peter Lang, [2018] | Includes bibliographical references.
Identifiers: LCCN 2017055795| ISBN 9783631744789 (Print : alk. paper) | ISBN 9783631744796 (E-PDF) | ISBN 9783631744802 (EPUB) | ISBN 9783631744819 (MOBI)
Subjects: LCSH: Energy security--European Union countries. | Energy security--Turkey. | Natural gas pipelines--Europe. | Natural gas pipelines--Turkey.
Classification: LCC HD9502.E852 T35 2018 | DDC 333.79094--dc23 LC record available at https://lccn.loc.gov/2017055795

Cover image: © chocolatefather / Fotolia.com
Cover Design: © Olaf Gloeckler, Atelier Platen, Friedberg

ISBN 978-3-631-74478-9 (Print) · E-ISBN 978-3-631-74479-6 (E-PDF)
E-ISBN 978-3-631-74480-2 (EPUB) · E-ISBN 978-3-631-74481-9 (MOBI)
DOI 10.3726/b13228

© Peter Lang GmbH
Internationaler Verlag der Wissenschaften
Berlin 2018
All rights reserved.

Peter Lang – Berlin · Bern · Bruxelles · New York ·Oxford · Warszawa · Wien

All parts of this publication are protected by copyright. Any utilisation outside the strict limits of the copyright law, without the permission of the publisher, is forbidden and liable to prosecution. This applies in particular to reproductions, translations, microfilming, and storage and processing in electronic retrieval systems.

This publication has been peer reviewed.

www.peterlang.com

The views expressed are purely those of the author and may not in any circumstances be regarded as stating an official position of Medipol University. **This book was supported by Medipol University Department of Political Science and Public Administration**

Table of Contents

Abstract 11

Zusammenfassung 13

Acknowledgements 15

List of Abbreviations 17

List of Figures 19

1. Introduction 21
   1.1 Purpose of the Book 21
   1.2 Literature Review and Argumentation 23
   1.3 The Structure and Methodology 25

2. Theoretical Framework: Interdependence Theory 27
   2.1 Energy Security and Interdependence 27
   2.2 Emergence of Interdependence Theory 28
   2.3 Central Concept; Introduction to Theory 29
   2.4 Types of Interdependence 31
        2.4.1 Vulnerability and Sensitivity 31
        2.4.2 Direct and Indirect Interdependence 33
        2.4.3. The Impacts of International Policy 34
        2.4.4. Actors 36
               2.4.4.1 Reducing Importance of Military Power 37
               2.4.4.2 Economic Power 39
               2.4.4.3 In Terms of Security Vulnerability and Sensitivity 40
   2.5. Complex Interdependence 42
   2.6 Compatibility with Realism and Complex Interdependence 42

3. Energy Interdependence as a Challenge ... 47
3.1 Pipeline Transit State on View of Vulnerability ... 48
3.2 Role of the Actors in Gas Policy ... 50
3.3 Diversification of Natural Gas Supply for the EU and Turkey ... 52
3.4 Cooperating for Energy Security Between the EU and Turkey: An Interdependence Approach ... 54
3.5 Contributing Energy Cooperation of Turkey to EU Membership ... 57
Summary ... 61

4. Global Growing Energy Demand ... 63
4.1 Increasing Importance of Gas ... 63
4.2 The Fossil Energy Resources used in the World and their Consumption ... 68
4.3 The Fluctuations in Energy Policy After 9/11 ... 72
4.4 Rise of Natural Gas Use in the Electric Power Sector around the World ... 74
4.5 Scenario of World Natural Gas ... 75
4.6 Competition between Actors on Energy Issues ... 77

5. EU Debates on Reducing Gas Dependency and EU-Russia Relations ... 81
5.1 Development of European Energy Policy ... 81
    5.1.1 After the Oil Crisis ... 83
    5.1.2 European Union Energy Policies Between 1986 and 1995 ... 85
    5.1.3 Energy Framework Programme ... 86
    5.1.4 White Paper: An Energy Policy for Europe ... 89

5.2 The Green Paper (The New Energy Policy of the EU) ... 89
    5.2.1 The Green Paper: A European Strategy for the Security of Energy Supply ... 89
    5.2.2 Sustainable, Competitive and Secure Energy for European Strategy (2006) ... 90

5.3 Europe Energy Charter and Europe Energy Charter Treaty ... 93

5.4. Security Risks and EU Policies on European Energy Supply 95
5.5 The EU's External Dependence and Energy Consumption 98
    5.5.1. The EU and the Security of Natural Gas Supply 100
    5.5.2. The Effects of the Gas Crises on European Energy Security ... 102
5.6 Energy Targets and Prospects for the EU in 2020 105
Summary 106
5.7. Analysis of the Cooperation between the EU and Russia on Energy Policy 108
    5.7.1 Russian Gas Pipelines and Projects to Europe 112
    5.7.2 South Stream 113
    5.7.3 Nord Stream Pipeline 115
    5.7.4 Gazprom within the State 116
Summary 117
5.8 Alternative Policy Options for Reducing Gas Dependency 119
    5.8.1 Promoting Use of EU Renewable Energy Sources 119
    5.8.2 Improving Energy Efficiency 121
    5.8.3 Nuclear Energy Use and Debate 123
    5.8.4 Single Energy Market and Liberalisation of the Gas Market in the European Union 124

6. Turkey's Energy Dependency and Role in the Southern Gas Corridor 129

6.1 Rising natural gas use in Turkey 131
6.2 Turkey's Dependency on Russia with Blue Stream Natural Gas Pipeline 134
6.3 Transmission of Gas Policy in Turkey 134
6.4 Turkey-EU Coherence after Negotiations in the Field of Gas Policy 136
6.5 Trans-European Energy Networks (TEN-E) 139
6.6 Aims of Turkey's Gas Policy by 2023 139
6.7 Turkey's Energy Chapter in EU Membership 141
    6.7.1. Renewable Energy Resources in Turkey 141
    6.7.2. Nuclear Energy Studies 142
    6.7.3. Energy Efficiency Studies 143

      6.7.4   Climate Change Policy and Position of Turkey in the
Kyoto Protocol 144
      6.7.5   Challenges of Turkey's Energy Policy and
EU Accession Process 145

  6.8   International Pipeline Projects as Energy Hub of Turkey 148
      6.8.1. Baku–Tbilisi–Ceyhan (BTC) 148
      6.8.2. The Samsun–Ceyhan Pipeline 150
      6.8.3. Iraq–Turkey Crude Oil Pipeline (Kirkuk–Yumurtalik) 151

  6.9   Important Export States 152
      6.9.1   Iran 152
      6.9.2   Turkmenistan 153
      6.9.3   Iraq 154

  6.10 Southern Gas Corridor 156
      6.10.1 Azerbaijani-Turkish Project Trans-Anatolian Pipeline 158
      6.10.2 Competing Existing Pipelines around Turkey's Border 161
             6.10.2.1 Short Version of Nabucco 161
             6.10.2.2 Trans Adriatic Pipeline 163
             6.10.2.3 South East Europe Pipeline 165
      6.10.3 Baku–Tbilisi–Erzurum 165
      6.10.4 South Europe Gas Ring Interconnector Turkey–Greece–
Italy Pipeline 166
      6.10.5 Azerbaijani-Turkish Natural Gas Pipeline Project (Shah
Sea I and II) 166

7. Energy as a Factor in the Relationship between Turkey and the EU 169

  7.1   Turkey's Geostrategical Factor 169

  7.2   Rising Importance of Turkey in European Energy Security 171

  7.3   Interest and Cost of Cooperation on the Southern Gas Corridor 173

  7.4   Changes to Turkey's Position 176

8. Conclusion 179

9. References 187

Annex: Interviews 203

Abstract

This study analyses the issue of energy security and natural gas in the Southern Gas Corridor, and its effects on the relationship between the European Union (EU) and Turkey. The usage of natural gas as an energy source has increased in recent years, and is expected to rise in the following decades. This increase in the consumption of natural gas has rendered itself of greater strategic importance in the world. Additionally, the conflict between Ukraine, Belarus and Russia over natural gas has threatened both the short- and long-term security of the EU's gas supply. Consequently, this conflict has created the need to find alternative routes in order to meet Europe's increasing gas demand. In this context, Turkey's and the EU's dependency on Russian natural gas has increased in the last decades, causing both Turkey and the EU to search for outside energy sources. In this respect, Turkey plays a key role as part of the strategic Southern Gas Corridor initiative, which is a proposed gas pipeline in cooperation with the Trans Anatolia Natural Gas Pipeline (TANAP) project, proposed to run from the Caspian region through Turkey to Europe. This study argues that the energy dependency among the countries not only affects economic decisions, but also political decisions. In this regard, it is claimed that the joint projects could have an impact on the behaviours and foreign policies of states. Thus, the relations are explained with the theory of interdependence taking into account the transit role of Turkey within the framework of gas import of the EU.

During the years from 1959 to 2000, energy, and in particular gas, did not play a significant role in Turkey–EU relations. Energy lines followed the routes outside Turkey and no significant crisis was experienced with Russia during this era. However, with the beginning of the new millennium it is claimed that both the political crises that hit Russia and the boost of gas consumption significantly increased the importance of Turkey. This importance had a substantial role in the EU's decision to grant Turkey the status of candidate country for EU membership in 2004. It is expected that in the 21st century, the role played by gas will increase in global energy security and thus the Southern Gas Corridor will be an essential alternative for Turkey and the EU. Therefore, this study argues that international joint projects, like these, are likely to affect the political decisions of Turkey and the EU. This study suggests that joint projects such as Baku–Tbilisi–Ceyhan, Baku–Tbilisi–Erzurum, the Turkey–Italy–Greece pipeline, the Kirkuk–Ceyhan pipeline, and finally the TANAP will accelerate Turkey's EU accession process. Cooperation also brings together the societies and increases and deepens diplomatic relations.

In this regard, with the opening of the Southern Gas Corridor, interdependency between the EU and Turkey will increase. With the opening of the Shah Deniz I and II gas fields to the EU market, other pipeline projects will accelerate. As it is put by the actors and EU bureaucrats, it seems hard for the EU to become a global power in the Middle East without Turkey. Taking into account the coming scenarios, the importance of Turkey will increase parallel to the increase in gas consumption in the EU in the near future. In particular, in addition to the opening of the Southern Gas Corridor, with its improved relations with the Northern Iraq regional administration, Turkey's position as a transit country for oil and gas resources to EU countries will also strengthen.

Zusammenfassung

Diese Studie analysiert die Zusammenhänge von Energiesicherheit und natürlichem Gas im südlichen Gaskorridor und ihre Auswirkungen auf die Beziehungen zwischen Europäischer Union und der Türkei. Die Nutzung von Erdgas in der Welt als Energiequelle hat in den letzten Jahren zugenommen, und es wird erwartet, dass sie in den folgenden Jahrzehnten steigen wird. Dieser ansteigende Erdgasverbrauch hat in der Welt größere strategische Bedeutung erlangt. Darüber hinaus hat der Erdgas-Konflikt zwischen der Ukraine, Weißrussland und Russland sowohl die kurz- als auch die langfristige Sicherheit der Gasversorgung der EU bedroht. Folglich hat dieser Konflikt die Notwendigkeit hervorgebracht, alternative Routen zu finden, um Europas steigenden Gasbedarf zu decken. In diesem Zusammenhang sei erwähnt, dass die Abhängigkeit der Türkei und der EU von russischem Erdgas in den letzten Jahrzehnten zugenommen hat, was die Türkei und die EU dazu veranlasst, nach auswärtigen Energiequellen zu suchen.

In diesem Zusammenhang spielt die Türkei eine Schlüsselrolle in Form von einer Teilhabe der strategischen Süd-Gas- Koridor Initiative. Dies ist eine Pipeline, welche in Zusammenarbeit mit der (TANAP) die Region um das kaspische Meer durch die Türkei mit Europa verbindet. Diese Studie behauptet, dass die Energieabhängigkeit unter den Ländern nicht nur die wirtschaftlichen, sondern auch die politischen Entscheidungen beeinflusst. In diesem Zusammenhang wird argumentiert, dass die gemeinsamen Projekte einen Einfluss auf das Verhalten und die Außenpolitik der Staaten haben könnten. Demnach werden die Beziehungen mit der Theorie der Interdependenz unter Berücksichtigung des Transit-Rolle der Türkei im Rahmen des Gasimports der EU erklärt.

In den Jahren 1959 bis 2000 spielte Energie, insbesondere Gas, keine bedeutende Rolle in den Beziehungen zwischen der Türkei und der EU. Die Energiestrecken folgten den Routen außerhalb der Türkei, und es gab während dieser Zeit keine signifikante Krise mit Russland. Mit dem Beginn des neuen Jahrtausends wird jedoch behauptet, dass sowohl die politischen Krisen, die Russland trafen, als auch die Erhöhung des Gasverbrauches die Bedeutung der Türkei deutlich erhöhten. Diese Bedeutung, die sich verstärkt hat, spielte eine wesentliche Rolle bei der Entscheidung der EU, der Türkei den Status eines Beitrittskandidaten für die EU- Mitgliedschaft im Jahr 2004 zuzugestehen.

Es wird erwartet, dass im 21. Jahrhundert die Rolle, die Gas spielt, die globale Energiesicherheit steigern wird, und damit wäre der südliche Gaskorridor eine wichtige Alternative für die Türkei und die EU. Daher vertritt die Studie die Auffassung, dass internationale gemeinsame Projekte wie diese sich wahrscheinlich auf die politischen Entscheidungen der Türkei und der EU auswirken. Diese Studie legt nahe, dass die gemeinsamen Projekte wie Baku-Tiflis-Ceyhan, Baku-Tiflis-Erzurum, die Türkei-Italien-Griechenland-Pipeline, die Kirkuk-Ceyhan-Pipeline und schließlich die TANAP den EU-Beitrittsprozess der Türkei beschleunigen wird. Die Zusammenarbeit bringt auch die Gesellschaften zusammen, verbessert und vertieft diplomatischen Beziehungen.In dieser Hinsicht wird sich mit der Eröffnung des südlichen Gaskorridors auch die Interdependenz zwischen der EU und der Türkei erhöhen. Mit der Öffnung der Shah Deniz I und II Gasfelder in Richtung EU-Markt, werden sich andere Pipeline-Projekte beschleunigen.

So, wie es von den Akteuren und EU-Bürokraten dargestellt wird, scheint es für die EU schwer zu sein, ohne die Türkei eine globale Macht im Nahen Osten zu werden. Unter Berücksichtigung der bevorstehenden Szenarien, wird sich in naher Zukunft die Bedeutung der Türkei parallel zur Erhöhung des Gasverbrauchs in der EU steigern. Zusätzlich zur Eröffnung des südlichen Gaskorridors, wird sich die Position der Türkei besonders mit seinen verbesserten Beziehungen zur Nordirakischen Regionalverwaltung, auch als Transitland für Öl- und Gasressourcen in die EU-Länder festigen.

Acknowledgements

I am sincerely and heartily grateful to my advisor, Prof. Dr. Otmar Höll for both his guidance, support and encouragement, and his invaluable insight and comments given to me throughout my study. I am sure it would have not been possible without his help

Lastly, I feel deeply indebted to my gracious wife Ferdane Esra TANRIKULU, along with my mother and father. Until the end of my life, I will remain grateful to them as they had to shoulder enormous burdens and make sacrifices throughout the creation of this book. I thus dedicate this thesis to them.

Faik TANRIKULU

List of Abbreviations

ALTENER	Programme for the Promotion of Renewable Energy Sources
BOTAS	Turkish State Pipeline Corporation
BTC	Baku–Tbilisi–Ceyhan Main Export Crude Oil Pipeline Project
BP	British Petroleum
BTE	Bakü–Tbilisi–Erzurum
Bcm/y	billion cubic metres per year
CO_2	Carbon dioxide
CPC	Caspian Pipeline Consortium
CFSP	Common Foreign and Security Policy
EC	European Community
ECT	Energy Charter Treaty
EMRA	Energy Market Regulation Authority EU – European Union
EUROMED	Euro-Mediterranean Partnership ECSC – European Coal and Steel Community ECS – Energy Charter Secretariat
EEC	European Economic Community
EURATOM	European Atomic Energy Community EAP – Environmental Action Programme
EAES	European Atomic Energy Society IEA – International Energy Agency
INOGATE	Interstate Oil and Gas Transport Programme
ITGI	Interconnector Turkey–Greece–Italy
LNG	Liquefied Natural Gas
MENR	Turkish Ministry of Energy and Natural Resources MFA – Turkish Ministry of Foreign Affairs
MTA	Mineral Research and Exploitation MTOE – Million tonnes of oil equivalent
MW	Megawatt
MWe	Megawatt electrical
MWth	Megawatt thermal
NPPs	Nuclear Power Plants
NSC	National Security Council of Turkey
NABUCCO	Turkey–Bulgaria–Romania–Hungary–Austria Natural Gas Pipeline OECD – Organisation of Economic Cooperation and Development
OPEC	Organisation of Petroleum Exporting Countries PJCCM – Police and Judicial Cooperation in Criminal Matters SPO – State Planning Organisation

SIE	Single European Act
SAVE	Specific Actions for Vigorous Energy Efficiency SOCAR – State Oil Company of Azerbaijan
SEEP	South East Europe Pipeline
TACIS	Technical Assistance to the Commonwealth of Independent States
TCGP	Trans-Caspian Natural Gas Pipeline Project
TEK	Turkish Electricity Authority
TETAS	Turkish Electricity Trading and Contracting Corporation TPAO – Turkish Petroleum Corporation
TACIS	Technical Assistance for Commonwealth of Independent States TEAS – Turkey's Electricity Generating and Transmission Corporation
TEN	Trans-European Network
TRACECA	Transport Corridor Europe Caucasus Asia
TUBITAK	The Scientific and Technological Research Council of Turkey
TUPRAS	Turkish Petroleum Refineries Corporation
TANAP	Trans Anatolian Pipeline
UNFCCC	United Nations Framework Convention on Climate Change
UNIDO	United Nations Industrial Development Organisation

List of Figures

Figure 1:	Gas Production and Consumption by Region	66
Figure 2:	World Energy Consumption	68
Figure 3:	Fossil Fuel Reserves Production (R/P) Ratios in 2011	70
Figure 4:	Rotterdam Product Prices and US Gulf Coast Product Prices	73
Figure 5:	World Electricity Production in 2008	74
Figure 6:	Primary Energy Demand in the New Policies Scenario	75
Figure 7:	Coal-fired Electricity Generation by Region in the New Policies Scenario	76
Figure 8:	EU-27 Energy Import dependency	99
Figure 9:	Energy dependency – natural gas, 2009	101
Figure 10:	Russian Exports	110
Figure 11:	Share of Russia's natural gas exports by destination, 2010	112
Figure 12:	Map of South Stream Pipeline Route Option	114
Figure 13:	Map of Nord Stream Pipeline Route Option	115
Figure 14:	Share of renewable energy in gross final energy consumption and target for 2020	120
Figure 15:	Installed capacity for electricity generation from renewables, EU-27	121
Figure 16:	Share of Nuclear in National Electricity Generation in 2009	123
Figure 17:	Turkey Total Primary Energy Supply	130
Figure 18:	Turkey's Natural Gas Imports	133
Figure 19:	Map of BTC Pipeline	149
Figure 20:	Map of the TANAP pipeline	160
Figure 21:	Map of Nabucco West Gas pipeline	162
Figure 22:	Map of the TAP Pipeline Route	163

1. Introduction

1.1 Purpose of the Book

The world is becoming a global village and the borders of nation states are beginning to blur through globalisation and interdependence in economics, politics and communications, as well as environmental issues. As a result, multiple channels of interaction, relationships among states and national sensitivity to outside developments have increased. In this sense, multiple channels such as international organisations, international trade, and national and international interest groups have started to take on important roles.

In today's world of globalisation, energy dependency affects not only the economic but also the political decisions of countries. The foreign dependency of countries in terms of gas and oil causes political crises from time to time between the exporting and importing countries. Therefore, these crises also change the decision of authorities and the foreign policies of countries. The dispute in gas prices experienced with Russia between 2006 and 2009, for example, turned into a political crisis over time and the EU Commission sought to find alternatives in its foreign policy as a result of this experience. Similarly, Turkey is concerned with the risk of Iran cutting the gas. Considering that this issue is being heard by an international court because of a dispute in gas prices between both countries, it seems to be possible that similar crises may be encountered in the future. Thus, both the EU and Turkey have to seek alternative possibilities in order to ensure their gas supply.

The reports prepared by and speeches made within the EU have recently spoken frequently about the role of Turkey in meeting the energy need and ensuring energy supply security. First, there has been a significant increase in the share of natural gas in global and European energy consumption. Taking into account the existing trends, it is estimated that the share of natural gas in the consumption will continue to increase and that there will be a dramatic decrease in the share of oil. This being the case, the pipeline route alternatives reaching through the EU will become more important. EU countries used to supply their energy needs from three fundamental sources, namely Russia, the Middle East and European countries. However, in recent periods, there has been a dramatic decrease in production of oil and natural gas sources in Norway and the UK. Therefore, the dependency of EU countries on the import of energy from Russia or the Middle East is gradually increasing.

Energy has become one of Turkey's priority policy fields in the last decade. In addition to the Baku–Tbilisi–Ceyhan, a further natural gas pipeline has been constructed between Turkey and Azerbaijan. The natural gas that comes from Interconnector Turkey–Greece–Italy (ITGI pipeline) is not only marketed to Turkey but also to Greece via Turkey. Not being content with these existing lines, Turkey continues with works for new lines. These works are not only conducted through the state, but also through the private sector. Turkish companies operating in Northern Iraq demonstrate great efforts in delivering the gas of Northern Iraq to Turkey, and from Turkey to the entire West. If this becomes a success, a new line will be added to the Kirkuk–Yumurtalik pipeline and the political and economic balances in the region will change. However, the increase in Turkey's role for the EU concerning these energy routes is not solely related to its geographic location. The remarkable progress in Turkey's economy in recent years, its experiences in the fields of energy and energy transportation, its sufficient manpower, and its political, economic and social structure—which is more stable compared to those of other countries in the region—further highlight the importance of the Turkish route. It is claimed that it is difficult to find a second country in the region that could provide such a reliable and quality service. As already mentioned above, the interdependency in energy has become a decisive factor in bilateral relations and changes to foreign policies and attitudes among the governments. In this context this dissertation will seek answers to following questions: To what extent do the EU and Turkey need each other in the field of energy in strategic terms, and why do they need a cooperation? What will change in EU-Turkey relations with the opening of the Southern Gas Corridor?

In this dissertation, it is argued that the joint projects affect the political decisions of states over time and increase interdependency. It is claimed that the importance of gas will increase in global energy security in the 21st century. Following the problems experienced between the EU and Russia and between Turkey and Iran, the increasing energy consumption of both the EU and Turkey demonstrates the strategic importance of Turkey as an energy hub.

Due to the fact that the EU energy consumption of oil and gas was low between 1950 and 2000, and that no significant crisis was encountered, energy had no impact on EU-Turkey relations. However, initially with the Baku–Tbilisi–Erzurum pipeline and later with the Baku–Tbilisi–Ceyhan pipeline and then with the Trans Anatolian pipeline (TANAP) project, Turkey has taken significant steps towards being an energy bridge. In line with these projects, Turkey–EU relations sped up and the accession process was accelerated. Meanwhile, the opening of the Southern Gas Corridor will also accelerate other pipeline projects, respond to Turkey's demand for energy and also guarantee energy security in the EU. In this sense, with the opening of the Southern Gas Corridor the interdependency between the EU and Turkey will increase and, thanks to the importance of Turkey in terms of its energy, it will become easier for Turkey to become an EU member as a result of the energy dependency of Germany and France in the coming 20 years, as stated by the Commissioner of Energy, Mr. Oettinger.

1.2 Literature Review and Argumentation

Turkey-EU relations are generally evaluated in other articles and studies. In this study, the focus is on the Southern Gas Corridor and its effects upon EU-Turkey relations. The process began with the Nabucco project, which hoped to open the Southern Gas Corridor. With the failure of the Nabucco project, it was claimed by a researcher that the Russian monopoly will continue. However, this study argues the contrary; the TANAP project developed an alternative policy which has been given the possibility to open the Southern Gas Corridor and will decrease dependency on Russia. As claimed in this study, the states have common economic benefits and mutual dependence, which minimises the problems and develops diplomatic relations between countries.

Interdependence theory claims that the effects of crises become less significant if countries of the world work together, and emphasises that the political behaviour can change over time. After 9/11, energy prices between 2003 and 2012 were doubled. After this attack, political tensions between countries became clearly evident. The Iraq war, the Arab Spring, tsunamis and lately the Libya crisis have resulted in disruption to the oil supply. The political struggle between Russia and Ukraine has shown the importance of the energy crisis. The priorities of foreign policy have started to change with energy policy and resulted in the development of alternative policy.

Naturally, the increasing gas demand of the EU and Turkey has brought regional cooperation. Being dependent on outside energy threatens both the EU and Turkey. Common energy projects such as BTC, BTE and lately TANAP will make the regions politically closer to each other.

My main hypotheses are as follows:

- With the opening of the Southern Gas Corridor, Turkey's strategic importance and its position as an energy hub will be strengthened.
- With the opening of the Southern Gas Corridor, Turkey's dependence on Russia and Iran for gas will decrease and it will be possible for it to diversify its energy supply.

- Political crises are experienced with the EU and the countries from where Turkey imports gas. If relevant precautions and alternatives are not designed, it is estimated that these crises will continue and affect the supply of energy
- The gas and oil reserves in Northern Iraq will change the balances in the region. As a result of the good relations maintained by Turkey with the Iraqi administration in recent years, significant steps will be taken in terms of transferring natural resources to Turkey and to the EU.
- BTE, BTC and recently the TANAP-TAP projects will significantly contribute to Turkey's accession to the EU.
- EU dependence on Russia for gas may restrict or hinder the EU's foreign policy in the future.
- The strategic importance of Turkey has increased in the aftermath of the 2000s. The joint projects to be carried out affect the political behaviours of the states. These projects will accelerate the accession negotiations.

When the study is considered as theory, with the first oil crisis in the 1970s, it was seen how countries were shaken up by the political and economic crises in the world. The crisis between Israel and Palestine ended with the embargo by oil exporting countries. Therefore, as stated in interdependence theory, the war and the internal crisis occurring in one place shake all other countries. Becoming an alternative to realist theory after World War II, the interdependence theory can be considered as a reflection of theory to reality. Meanwhile, it is only in the realistic approach that it is seen that military strength is not sufficient for the welfare of the country. On the other hand, in the interdependence theory approach, purely military methods are not rejected, but at the same time the increase in economic cooperation is realised to be more useful for the international affairs of the countries. As explained in this study, when considering Turkey's hands-off attitude with Northern Iraq and the experienced political crisis, this resulted in the loss of both sides. By understanding that this tension is damaging the benefits of both sides, Turkey, the central government and also Northern Iraq have recently redesigned their foreign policy. I present as an argument in this thesis that the change in foreign policies of the countries is obviously seen by the mutual benefits. In this regard, gas and oil agreements between Turkey and Northern Iraq are accepted as a milestone by the EU and Turkey. Meanwhile, the acceleration of the EU negotiation for accession in parallel with Turkey's steps toward being an energy bridge such as BTC, BTE and TANAP can be associated with this theory.

It has been seen that the economic benefits of the countries can affect their political decisions. The interdependence theory defends removing the trade barriers and increasing the regional cooperation. In this regard, the relationship with partner countries is affected negatively in parallel with military expansion. These observed problems can create economic sanctions. Thus, not giving up its nuclear programme caused economic sanctions for Iran. Recently, the decreasing uranium enrichment of Iran has led to a positive effect in the market. So, it is clearly understood that the reduction of political crisis has an affect on energy prices and foreign policy.

1.3 The Structure and Methodology

The methodology of this study has two aspects. First, the theoretical framework is based on a review of the literature on interdependence theory, which explores mutual cooperation and dependence within Turkey's role in the Southern Gas Corridor initiative. The second aspect of the methodology is empirical research, statistics, figures, interviews and tables, which aim to measure and demonstrate these three proposed forms of mutual cooperation on natural gas supply and sources between Turkey and the EU.

In order to understand the mutual dependence and cooperation between the EU and Turkey with regard to the Southern Gas Corridor, the theoretical framework on inderdependence theory is outlined in the second chapter. In the third chapter, Turkey's role in the Southern Gas Corridor within the framework of interdependence theory is explained. In the following section, the member states of the EU and the general situation of world gas policy are reviewed. The historical development of the policy is also presented in order to reach a better understanding of the background of today's developments. After that, an examination of the EU's energy policy is conducted by considering the three pillars of the policy, namely the common market, security of energy supply and environmental protection, and emphasised relations between the EU–Russia/Caspian in the energy field. This study aims to explore the current literature on energy security, in particular natural gas geopolitics within the Russia–Turkey–European Union triangle. In the following section, key energy sub-sectors in Turkey are concentrated on by underlining the energy situation in Turkey using appropriate quantitative data, such as recent data on primary gas energy consumption, and Turkey's production and energy demand, which is illustrated by tables and figures. Turkey's high dependency on the primary energy resources of oil and natural gas—almost all of which are imported—is the subject important in this section.

In the final section, energy is taken as one of Turkey's most significant trump cards in its quest for full membership in the EU. In the results section of this dissertation, the importance of the development of the Southern Gas Corridor to meet future energy demand in Europe is then researched, followed by a section which analyses what is meant by an energy transit state and looks at Turkey's track record as a state with regard to gas. Turkey's potential role in the development of the Southern Gas Corridor is then highlighted.

2. Theoretical Framework: Interdependence Theory

2.1 Energy Security and Interdependence

At the beginning of the 1990s, energy security in the theory of international relations emerged thanks to Barry Buzzan in the Copenhagen School. The term energy security is used in a military context as well as in exclusive connection with oil and gas resources. Today there is no longer a general concept of energy security. For some countries, energy security means depending on imports, and for others, exporting energy resources.[1] At this point, energy resources are in the fields of the military, politics and also industry. On the other hand, for a transit state, energy security means a source of income for industry and investment.[2] Additionally, there are several different terms used in relation to energy security. Sometimes this term is used by importers and consumers as *"energy security"* or often terms like *"security of supply"*, while energy security is limited to the aspect of the quantitative sufficiency of energy resources for energy supply. In such cases energy security means that a risk is arising due to domestic or regional political instability.[3] Robert Keohane and Joseph Nye first raised the concept of interdependence in their book *"Power and Interdependence"*. The definition of interdependence is one of mutual dependence. According to Keohane and Nye, interdependence was used simply as an *"interconnectedness"* of cost. Here, connectedness simply refers to the amount and frequency of interaction between the parties. Interdependence is assumed based on the political significance of this interaction and also the association with the development of costs, such as restriction of autonomy.[4]

---

1 **Ostry, Sylvia (2006)**; Sustainable Development and Energy Security: The WTO and the Energy Charter Treaty, Paper, Moscow, p. 13.
2 **Spelmenn, Scott (2007)**; The NATO School Energy Security Conference, Issue Paper Nr. 3, p. 2.
3 **Umbach, Frank (2003)**; Globale Energiesicherheit, (Global Energy Security) Oldenbourg Verlag, München, p. 51.
4 **Spindler, Manuela (2006)**; Interdependenz, in: Schieder, Siegfried / Spindler, Manuela (Hrsg.), Theorien der Internationalen Beziehungen, (International Relations Theories) 2. Auflage, Opladen, p. 96.

2.2 Emergence of Interdependence Theory

David Baldwin, Richard Cooper, Robert Keohane, Joseph Nye and Richard Rosecrance are the main contributors to the interdependence assumption. They argue that interdependence theory considers the mutual dependence of the politics and processes within government and acts on these systems according to outside conditions such as foreign policy decisions and behaviours.[5]

When we consider the roots of interdependence theory, Reinhard Meyer argues that the interdependency approach is based on ideas of liberal economics from the 18th century and developed critical analyses from basic ideas of mercantilism. In this way, the concept of interdependence is associated with the idealistic vision of international relations. He argues that by consistently following the principles of free trade, a world society, a *"world of interdependences"* could be created. In this respect, international division of labour was developed by the help of free trade on all other economic actors. This created interdependence promotes the harmonisation of interests and reduces conflict. These ideas are based on both the Manchester liberalism of the 19th century and the federalism of the 20th century.[6]

The liberal or Grotian tradition tends to point out the effects of domestic and international societies, and interdependence and international institutions. Later, liberal theories lost their significance. However, there was sharp opposition between realist and liberal theories. As sophisticated versions of liberal theory emphasise, *"the way interactions among states and the development of international norms interface with the domestic politics of the states in an international system can transform how those states feature their interests. In this case, transnational as well as interstate interactions and norms lead to new definitions of interests as well as new coalition possibilities for different interests within states".*[7] In the four industrialised countries of the USA, Japan, Germany and France, the end of the oil crises in 1974 was met with the attempt to confront the crises phenomena efforts with a global economic coordination process. As a result, the World Economic Forum was established after the oil crises of 1975. Manuel Spindler believes that politicians are responsible for their voters, and they should consider adjustment and stabilisation processes. Political rhetoric has changed the national governments of

---

5 **Lehmkuhl, Dr. Ursula (2001);** Lehr und Handbücher der Politikwissenschaft: Lehrmkuhl, Theorien Internationaler Politik (International Policy Theories), Oldenburg, 3. Auflage, p. 193.
6 **Ibid,;** p. 194.
7 **Keohane, O. Robert/S. Nye, Joseph (1989);** Power and Interdependence, Harper Collins Publishers, Preface to Second Edition, xi.

Western states, for example while combatting crises and global mutual dependencies. Therefore, international coordination is needed for the global economy in order to ease the implementation of painful economic measures for society. For this reason, interdependence theory has emerged in political science.[8]

Moreover, Keohane and Nye attempted to explain the patterns of change within aspects of the realist and liberal traditions during the mid 1970s in their book. They argue that *"reducing the United States' vulnerability to external shocks could be part of a strategy of policy coordination and international leadership"* and they stress, on the other hand, that the American oil stockpile is leading in the International Energy Agency and also US has supported international energy policy itself since the 1970s

In this case, they argued for effective international policy coordination on global issues that have been debated around the world, such as ecological and energy issues, and climate change. They believe it could be difficult to coordinate on these issues but that these ecological, energy, and climate problems could be solved through collective leadership.[9]

2.3 Central Concept; Introduction to Theory

As mentioned above, the interdependence theory is characterised primarily through international free trade and a limited natural resources policy by mutual interdependent relationships. This theory resulted from the rising integration of world trade. In the 1970s and 1980s, the notion of interdependence had become a central category of primarily empirical-analytical approaches to the study of international politics.[10]

Keohane and Nye advocate that *"interdependence has an effect on world politics and the behavior of states and also governmental behavior influence patterns of interdependence. Thus, governments regulate rules or institutions for certain activity, control transnational and interstate relations."*[11] Joseph Nye and Robert Keohane cited the following regarding interdependence: *"where there are reciprocal costly*

---

8 **Siegfried, Schieder/ Manuela Spindler (2006);** Theorien der Internationalen Beziehungen (International RelationTheories) (Hrsg.) Opladen Hill, p. 95.
9 **Keohane, O. Robert/S. Nye, Joseph (1989);** Power and Interdependence, Harper Collins Publishers, Preface to Second Edition, xii.
10 **Siegfried, Schieder/ Manuela Spindler (2006);** Theorien der Internationalen Beziehungen (Hrsg.) Opladen Hill, p. 107.
11 **Keohane, O. Robert/S. Nye, Joseph (2001);** Power and Interdependence, 3rd edition Longman, p. 7.

effects of transactions, there is Interdependence." Furthermore, they highlight that *"the interdependent relations will always involve costs, since interdependence restricts autonomy; but it is impossible to specify a priority whether the benefits of a relationship will exceed the costs."*[12] Apart from this definition, according to Stubbs, interdependence means that *"different patterns of interdependence affect patterns of cooperation, creating a conflict in the international system".*[13]

On other hand, Manuela Spindler states that after September 11th, sovereign internal politics are no longer in existence. In this sense, she means that after 9/11, the world has become more interdependent, which is not without effects on other countries in the events and decisions for policy and economy, for example, the Asia crises. Asian financial crises can have a significant impact on the European, North American and Latin American economies, ultimately requiring the world to work together to reduce the impact of these crises. This kind of causal relationship in international politics is often included in the concept of interdependence and is, in fact, the phenomenon that is known as interdependence. It has long been part of economic considerations in the framework of classical theory, which focuses its attention on international interdependencies in the field of international trade and monetary policy. Their ideological roots are in the classics of the free trade theory and political liberalism, ideas attributed to Adam Smith, David Ricardo and John Stuart Mill.[14]

The first conceptualisation of interdependence in political science emerged from Keohane and Nye, particularly with their book *"Power and Interdependence: World politics in transition"*, published in 1977. This book changed the structures of the international system, which exhibits a restriction on the capacity of states and loss of politics with consequences for the achievement of national economic and political goals because of mutual dependencies. According to Manuela Spindler, there are two possibilities for action of national policy under these conditions, after appropriate policy instruments have responded to these changes:

- Foreign policy–foreign policy trade
- International policy–cooperation between countries.[15]

---

12 **Ibid.;** p. 9.
13 **Stubbs, Richaerd/Underfhill, Geoffrey (2006);** Political Economy and the Changing Global Order:2006, Oxford University Press, p. 7.
14 **Siegfried, Schieder/ Manuela Spindler (2006);** Theorien der Internationalen Beziehungen (Hrsg.) Opladen Hill, p. 93.
15 **Siegfried, Schieder/ Manuela Spindler (2006);** Theorien der Internationalen Beziehungen (International Relation Theories) (Hrsg.) Opladen Hill, p. 97.

Effectively achieving the goals of the individual states and making the decisions required to do so depends on the interdependent relationship of all connected states; this process is defined as the collective action problem. Due to the unilateral pursuit of the objective, the creation or distribution of goods such as economic wealth, security, environmental protection for all states must be held as optimal. Therefore, for Keohane and Nye, interdependence conducts, under certain conditions, the attention of rational egoists acting in cooperation.[16]

In summary, interdependence theory advocates a multi-dimensional web of relations between states, their societies and inter-governmental organisations. In this sense, interdependence is understood as mutual, reciprocal dependence, but there is a distinction from simple interconnectedness.

2.4 Types of Interdependence

2.4.1 Vulnerability and Sensitivity

There are two measures for understanding the role of power in the interdependent relation: sensitivity and vulnerability. *"The sensitivity implies degrees of responsiveness within a policy framework of the politics, that is to say how quickly do the changes in one country bring costly changes in another? Sensitivity Interdependence can be defined as social, political or economic."*[17] It is not just about the costs that caused a failure or a change of trade and financial flows between countries directly, but also the changes of transactions in companies and governments. In this regard, sensitivity in systems is related to social, political and economic levels and their interactions with each other, not only by constant "framework of policy" acts. Interdependent relationships are triggered by a party's changes when the other has to carry out policy changes in their own interest as quickly as possible in order to avoid further costs.[18]

Keohane and Nye use the oil crisis and its effects on Japan, Western Europe and the United States in 1971 as an example of sensitivity interdependence, stating that *"sensitivity of these economies was function of the greater costs of foreign oil and the proportion of petroleum price rises."*[19] The United States in particular was sensitive to the outside change. Another example of sensitivity interdependence is the International Monetary Fund (IMF). Keohane and Nye added *"the*

---

16 **Ibid,:** p. 107.
17 **Keohane, O. Robert/S. Nye, Joseph (1977);** Power and Interdependence: World Politics in Transition, Brown Company, Boston, p. 12.
18 **Ibid.;** p 13.
19 **Ibid.;** p. 13.

example of European governments as being sensitive to changes in monetary policy and additionally, the United States as being sensitive to European decisions in the conversion of dollars into gold."[20]

If we consider sensitivity interdependence in social, political and economic issues, we see that development of radical student movements increased during the late 1960s by knowledge of each other's activities. Writers argued that the growth of transnational communication has provided sensitivity. In the following, they point out that the United States and Great Britain were affected under the Bretton Woods monetary regime in 1960 and were sensitive to foreign speculators or central banks, whereas the United States were more vulnerable than Britain because of their option on changing the rules.[21]

On the other hand, vulnerability can be explained as *"an actor's liability to suffer costs imposed by external events even after policies have been altered. The framework of policy could be altered and very different policies would be possible, which would be the cost of adjusting to the outside change".*[22]

Power potential and possibilities for action that arise under these conditions materialise from the fact that it is in the individual policy areas, usually dealing with asymmetric interdependence. This means that different countries are vulnerable in different policy areas. Keohane and Nye state that interdependent vulnerability includes a strategic dimension that allows states in positions of relative invulnerability the possibility of manipulating the international system for their own self-interest; they will try to take advantage of asymmetric interdependence as a source of power and to influence international organisations.[23] It was also mentioned in Christiane Lemke's lecture that the world market dependence on energy supply distinguishes between vulnerability interdependence and sensitivity interdependence, and is a mutual interdependence. Hence, this interdependence makes states more vulnerable to situations such as economic dependency on raw material resources.[24]

---

20 **Siegfried, Schieder/ Manuela Spindler (2006);** Theorien der Internationalen Beziehungen (International Relation Theories) (Hrsg.) Opladen Hill, p. 94.
21 **Keohane, O. Robert/S. Nye, Joseph (1977);** Power and Interdependence: World Politics in Transition, Brown Company, Boston, p. 13.
22 **Ibid.;** p. 13.
23 **Siegfried, Schieder/ Manuela Spindler (2006);** Theorien der Internationalen Beziehungen (International Relation Theories (Hrsg.) Opladen Hill, p. 106.
24 **Lemke, Christiane (2001);** Internationale Beziehungen Grundkonzepte, Theorien und Problemfelder (International Relation Basic Concepts, Theories and Problem) 2. Auflage München, p. 24.

Vulnerability is particularly significant for understanding the political structure of interdependence relationships. In this respect, vulnerability is more relevant than sensitivity. They give examples for vulnerability interdependence by analysing the politics of raw materials, such as the considered transformation of power after 1973. It is stressed that vulnerability is used in sociopolitical and politico-economic relationships. *"In the late 1960s the vulnerability of societies to radical movements depended on abilities to adjust national policies to deal with the change and decrease the cost of disruption."*[25]

As a result of this approach, sensitivity does not necessarily lead to vulnerability, and a degree of the former need not be equal; for instance, when both countries are equally dependent on imports of raw materials. Contrarily, both parties possess a similar extent of sensitivity to an external action.[26]

2.4.2 Direct and Indirect Interdependence

Within this context, the notions of interdependence assume that states are connected to each other by mutually dependent relationships and make changes of a military, political and economic nature. These changes provide a state with advantages, but for other countries can also exhibit serious drawbacks. Richard Cooper has proposed that the economic costs and inflationary effects of such changes can be integral in price or interests relations.[27] Besides sensitivity and vulnerability, there are two other measures of interdependence approaches, which are direct and indirect interdependence for core analytic distinction.[28] After the Second World War, direct and indirect interdependence were separated. As a result of the process of block formation, the resulting structures and interaction processes of world politics were divided into direct and indirect interdependence relationships. As put by Lehmkuhl, direct interdependence means both the mutual dependence between two blocks and their representative superpowers in addition to mutual dependence between vertical superpowers. At the same time, direct interdependence points out horizontal and vertical structures. Indirect interdependence defines those cases in which two superpowers engage in a dispute

---

25 **Keohane, O. Robert/S. Nye, Joseph (1977)**; Power and Interdependence: World Politics in Transition, Brown Company, Boston, p. 15.
26 **Ibid.**; p. 10–11.
27 **Cooper, Richard (1968)**; The Economics of Interdependence in the Atlantic Company Mc Graw Hill Book Company, New York.
28 **Meyers, Reinhard (1979)**; Weltpolitik in Grundbegriffen, Ein lehr und ideengeschichtlicher Grundriss, (The World politic in main component) Droste Verlag, p. 295.

among their clients at a low-level system. Lehmkuhl gives examples for indirect interdependence, such as the Indochina Middle East War, which happened without the parties being directly involved in a conflict. To sum up, indirect vertical interdependence is illustrated for the identification of the relationship between one or two blocks and third world countries.[29]

2.4.3. The Impacts of International Policy

According to Ursula Lehmkuhl, interdependence research focuses on the analysis of political, economic and social interactions, and interrelationships. She distinguishes between the following four research areas within this approach:

- "*The act of linking:* Theory on the relationship between domestic and foreign policy (Rosenau 1969)
- *The Transnational Politic* (Kaiser 1969)
- *The research approach of economic interdependence* (Cooper 1968; Bergsten 1973)
- *Power and Interdependence* (Keohane/Nye, research on the concept of mutual interdependence by the perspective of asymmetries)"[30]

International relations' influence on the foreign policy of states and autonomous or semi-autonomous actors is related to operating transnational organisations. For instance, the Roman Catholic Church, trade associations, trade unions, revolutionary movements and multinational companies' autonomous or semi-autonomous organisations control vast resources and have their own policies and aims.

The other characteristic of interdependence is that world policy issues are currently changing. Traditional analysis argues that the agenda of world politics still constitutes power, military force and violence, but the significance of these components has steadily decreased and there is less political interest concerning these topics. Although traditional analysis still emphasises the significance of the crisis in disarmament negotiations of the Middle East and Asia crisis, since the beginning of 1970s, the law of the sea negotiations on international trade, international

---

29 **Lehmkuhl, Dr. Ursula (2001);** Lehr und Handbücher der Politikwissenschaft: Theorien Internationaler Politik (International Politic Theories), Oldenburg, 3. Auflage, p. 196.
30 **Lehmkuhl, Dr. Ursula (2001);** Lehr und Handbücher der Politikwissenschaft: Theorien Internationaler Politik (International Politic Theories), Oldenburg, 3. Auflage, p. 194.

drug trafficking, air piracy, and increasing oil prices have been on the agenda of international policy and determined international events and problems.[31]

Students from different countries within the framework of the global mass media, military officials and ethnic groups mutually influence each other's behaviour and policies. As a result of international politics, Nye and Keohane formulated five issues that arise from the effects of transnational interaction and organisations.

- *Change assumptions*
- *Develop transnational pluralism*
- *Dependence and mutual dependency of states*
- *States potentiality to influence other countries*
- *Autonomous organisations of their own policies, which can interfere with the foreign policies of states.*[32]

According to Nye and Keohane, transnational relations change the parameters of states' foreign policy. These relations are effective on foreign policy. Thereby, hundreds of people from different countries or their communication through the mass media, their observations, thoughts and perceptions can lead to change.[33] According to Nye and Keohane, the second effect of transnational relations leads to the development of international pluralism. In this process, national interest groups with a transnational structure, particularly those that perform activities similar to transnational organisations, are associated with international pluralism. In this context, international non-governmental organisations, which perform tasks similar to those of national interest groups, have increased steadily. The emergence of new national groups leads to a rise in transnational organisations and the internationalisation of domestic policy. However, transnational organisations and national interest groups think of how to realise their state and society's desires through transnational connections.[34]

In summary, transnational relations can occasionally influence international policy. On the other hand, transnational organisations can also have a positive effect on relations between states. For example, the American multi-national oil

---

31 **Sullivan, Michael (1989);** Transnationalism, Power Politics, and the Realities of the Present System "International Relations in the Twentieth Century, A Reader": **Marc Williams** (E.d) London: Mach Millan Education, p. 261.
32 **Nye, Jr. Joseph S./Robert O. Keohane (1972);** Transnational Relations and World Politics; An Introduction, Cambridge, Massachusetts Harvard University press, xvii–xxii.
33 **Ibid,;** xviii.
34 **Ibid,;** xix.

companies that operate in the Middle East have positive relations between states of region.[35]

2.4.4. Actors

With the increasing number of developed I-GO and NGOs, the varying but remarkable political power and some key developments simply cannot be explained by state-centric theories. As mentioned above, among these is the withdrawal of the United States from Vietnam, which is an example of how military intervention and decisive state activity against an opponent can be proven wrong. On the other hand, there are events such as the oil crisis, in which an economic cartel was able to influence world politics without military intervention. In this case, the interdependence theory in political sciences refers to independent social actors who are affected by one another's behaviour. Structural involvement and structural interdependence can be typified at different levels of abstraction. Despite the increase in the number of international organisations, states' activities are increasing at the same time. In the past, states' activities were neglected, however, states' activities are now evaluated.

For instance, before World War I, the activities of international monetary movements were not significant. However, increasing state activity makes the state more sensitive. Of course, all of those developments and the policies of pursued states have led to the emergence of mutual dependence. In this sense, state and government control has weakened, particularly due to the impact of technological and social developments, and the state cannot provide full control over such events. The final analysis of these developments and the increasing activities of transnational relations of states with a sense of control over the capacity of these events create a gap between states. This gap has made transnational relations among countries more advantageous in this regard, and in terms of control center.[36]

Keohane and Nye focused recently on the role of international organisations. According to them, there are a large number of international organisations in the communication channel. Therefore, the different and distinct roles of international organisations in world politics should be emphasised. On the other hand, Hans Morgenthau, a traditional realist, claims that states in which the governments have been struggling to achieve their own interests and power view security issues as the main problem. In such a world, international institutions are expected to have

---

35 **Nye, Jr. Joseph S./Robert O. Keohane (1972);** Transnational Relations and World Politics; An Introduction, Cambridge, Massachusetts Harvard University press, xxi–xx.
36 **Ibid,:** xxii.

a limited effect; the importance of international institutions in world politics is therefore marginal. In the framework of transnational governments and transnational coalition, increased political bargaining processes are part of the role of international organisations. These negotiations are now in the process of coalition building, and are performing a catalysing role in the international agenda of weak and strong states, as well as using political initiative; furthermore, they also play a role as a basis for strategy implementation.

These organisations, by identifying key issues and deciding which issues may take place in the same group, can support the determination of the priorities of governments, and other governmental activities. Under the 1972 Stockholm Environment Conference, environmental organisations were established, strengthening their position. In 1974, the World Food Conference, organised by the United Unions, took action against possible food needs. The United Nations began to discuss directing means towards the policy of third world countries within governments. The activities of governments, international organisations like the IMF and the GATT focused on direct investments and trade issues.[37]

According to Keohane and Nye, international organisations, which bring together officials, at the same time can aid the emergence of potential coalitions. Among poor states, the solidarity of the third world and strategies under supervision of the United Nations have been discussed and developed at international conferences. International organisations also provide opportunities for coalition with other non- governmental bodies or transnational organisations by using communication channels.[38]

At the same time, international organisations also provide rationale linking the strategies of small and weak states. During the New International Economic Organisation talks, third world states have attempted to address the increase in oil prices along with other issues.[39]

In summary, interdependence has a large number of channels that are connected to each other.

2.4.4.1 Reducing Importance of Military Power

Keohane and Nye point out that the importance of military power persists in world politics, such as in Vietnam, the Middle East and India, as well as in the

---

37 **Keohane, O. Robert/S. Nye, Joseph (1977)**; Power and Interdependence: World Politics in Transition, Brown Company, Boston, p. 35–36.
38 **Ibid,;** p. 36.
39 **Ibid,;** p. 36.

Pakistan wars, the Iraq war and the Soviet influence in Eastern Europe. Accordingly, mutually dependent states have decreased the probability of the applicability of military power. Military power conversely takes an important place in the relations of states that do not have interdependent relations. In such cases, interdependence could significantly influence the policy of other states.

As is well known, some political scientists emphasise the importance of military power in international politics. In both World Wars, security issues became the priority for every state; however, military power, supported by economic factors and other sources, was strengthened. When states' primary objective is survival, military force is seen as a central importance.[40]

According to Keohane and Nye, military power is not seen as an appropriate method for economic and ecological issues. However, among developed countries in economic matters, the use of force or threat is taken into account. In such cases, the assumptions of the realist school can be realised. In most cases, the use of military force's influence is uncertain and costly; however, the use of military force should be considered in some cases. Each of the superpowers during the Cold War attempted to use military force, with the aim of deterring attacks. These countries provided a deterrent force through the use of the military, and the protective roles of the problems between the allies used it as a bargaining tool. For example, the United States and European countries protected their needs in Europe, in the possession of military power, and the level of commercial and financial relations was used with these states.[41]

As Werner Levi emphasises in his work *"The Coming End of War"*, the national interest internationalises increasingly in the case of possible war, and he asserts that states can have more losses than profit: for example, he points out the use of nuclear power as damaging to the world because of the risk of nuclear crises and wars. The invasion of Afghanistan, the Iran–Iraq war, Iraq's occupation of Kuwait, American intervention in Iraq, the Bosnia and Herzegovina, Armenian–Azerbaijani and the Kosovo conflict are all important indicators that this type of war has not disappeared.[42]

East–West relations, as well as North–South relations, or the third world countries are important for power in mutual relations. The military power of the Soviet Union as well as its economic control over Eastern Europe played an important

---

40 **Ibid**; p. 27.
41 **Ibid**: p. 27–28.
42 **Sullivan, Michael (1989)**; Transnationalism Power Politics, and the Realities of the Present System "International Relations in the Twentieth Century, A Reader": **Marc Williams** (ed) London: Mach Millan Education, p. 268.

political role. American hidden or open use of force was seen in the Caribbean and particularly in Guatemala in 1954 and the Dominican Republic in 1965.[43]

2.4.4.2 Economic Power

From a realistic point of view, the military security of states is the main objective in situations that military force affects directly or even in issues regarding unrelated defense of the country. Military issues are considered not only of secondary importance, but are also evaluated at the same time in terms of politics and military. For instance, the balance of payments problem and the financial implications for the state are both taken into account. In this sense, in 1964 with McGeorge Bundy, the devaluation of the dollar should be considered only if it is necessary to fight in the Vietnam War. In the same way, Henry Fowler, US Secretary of Commerce in 1971, pointed out that the US, with regard to the defense of the West, would have to be effective to maintain their leadership and should be trading annually with a surplus of at least four to six billion USD.[44]

Keohane and Nye stress that sovereign states may strive in this time to use economic force over other matters. However, only economic objectives can give successful results when it comes to the use of economic power. The economic objectives are at the risk of causing political implications. International actors may be different according to subjects, and international organisations may change by negotiations. Using a strong military force as a tool against one another can lead to dangerous consequences. International organisations are able to use a link as an instrument for poor and weak states, which is much easier and less risky.[45] According to Keohane and Nye, if the role of military power is negligible, states may use other tools in term of interests and power. Within asymmetric dependence between states, power would be the main factor. States as well as international organisations and transnational actors and relations will want to use power for this purpose. States should take into consideration the impact on economic interdependence on the prosperity of citizens with regard to the power element. Economic and ecological interdependence often incurs large gains or large losses. Potential mutual losses and gains and the dangers of any deterioration of the status of the actor will encourage others to be careful in using asymmetric interdependence.[46]

---

43 **Keohane, O. Robert/S. Nye, Joseph (1977);** Power and Interdependence: World Politics in Transition, Brown Company, Boston, p. 28.
44 **Ibid:** p. 30.
45 **Ibid:** p. 31.
46 **Ibid;** p. 32.

Today, particular issues of non-military inter-state relations are generally quite important, while other issues can be neglected or evaluated on a technical level. International financial and trade issues, especially those related to the oil or food sectors or multi-national companies and financial crises, have been important issues in the last twenty years, yet are not on the agenda of international relations.[47] Keohane and Nye, in an analysis of traditional international policy, do not emphasise agenda issues. According to the traditional approach, military and security issues are seen as the main subjects of foreign policy in international politics and it is assumed that economic issues are classed as secondary level matters. Therefore, security issues are high policy, and economic issues are classed as low policy.

However, issues and actors in world politics have been varied and increased, reduced in the use of power, blurred between domestic and foreign policy; therefore, mutual dependency conditions can be roughly estimated, and differ from the agenda of international politics.

States may politicise the issues and can carry a variety of topics to the international agenda from an internal problem. According to the changes of the distribution of power, natural resources can also affect the international agenda. In the 1970s, oil- producing countries changed the international agenda completely due to high consumer demand and increasing interest from multinational corporations.[48]

2.4.4.3 In Terms of Security Vulnerability and Sensitivity

According to recent studies in the gas sector, since the attacks of September 11, there has been an increase in political and geopolitical issues. After the Russian-Ukrainian gas crisis of 2006, the potential for economic cooperation between stakeholders and representatives of political and geopolitical actors has increased. In this way, rational actors are trying to minimise the effects of the costs as much as possible, while at the same time, they strive to attain the highest possible advantages. Keohane and Nye distinguish the use of asymmetric interdependence from the use of interdependence sensitivities and vulnerabilities in both non- military as well as military contexts, and state that the use of asymmetry is both dominant in the military when compared to the non-military sector, and connected with higher costs. The vulnerability dimension refers to which of the actors have to

---

47 **Ibid;** p. 32.
48 **Keohane, O. Robert/S. Nye, Joseph (1977);** Power and Interdependence: World Politics in Transition, Brown Company, Boston, p. 33.

bear lower costs in the long term in order to make alternatives available. Thus, they are able to set up the rules in a relationship.[49]

Keohane and Nye state that a consideration of interdependence should pay attention to vulnerability, as there are always possible costs when it comes to adjustment of policy. It is difficult to implement changes to policy in the short term. The impact of external changes will mostly reflect the interdependence sensitivity. Vulnerability can only be measured by costs, which arise from certain periods in the adaption of changed environment. However, measurements of the immediate consequences of environmental change are not necessarily contingent on long-term sensitivity, but still less related to the long term.

Long-term interdependence vulnerability depends both on the political desire of an actor as well as their skills and capabilities. An example of this is given by Keohane and Nye who provide the case of two states. One state imports 40% of its crude oil, which is affected by price increases. If one state had access to relatively low-cost alternative energy sources, then the other state would be subject to vulnerability. To determine this vulnerability, it is important to determine which price is sufficient for the raw material or alternative raw material and can be obtained through a change of policy. A state's dependency on imports cannot be considered when determining vulnerability. A high import quota is more useful for determining high sensitivity when, for example, both states are experiencing price increases. Actors want to behave according to certain strategies and question what other opposing actors will cost or provide. In short, long-term strategy or politics are based on in-depth analysis and potential vulnerability.

Keohane and Nye also bring this to an example from the energy sector. An energy company demands a state's oil, but how much oil the state claims and to whom it sells the energy source is completely up to the state to decide. Thus, the concern of the state lies in the sensitivity level, as the quantity and price of oil affect a significant portion of government revenues and expenditures. This situation can cause high prices and a cease in production and, by increasing the state's cost too much, the state could be forced to trade. The state can attempt to switch to the next highest vulnerability level, which is more consolidated and performed in extreme cases, such as nationalising the company. Any attempt by the company to use its advantageous position in the sensitivity dimension carries a high probability of the group ending up with a deteriorating situation. Interdependence sensitivity can therefore only serve as a power source of influence when there is possibility. At this point, the influence of the asymmetry of a

---

49 **Keohane, Robert O. / Nye, Joseph (1977);** Power and Interdependence, p. 13.

derivative in the sensitivity dimension is very limited if underlying asymmetries are on a vulnerability level.[50]

2.5. Complex Interdependence

Keohane and Nye mention that communities, in their complex interdependence, have increased communications channels on domestic and international policy. This policy has brought uncertainties. According to this assumption, there is not any restriction on creating a coalition on the border of a country. The situation could lead to political bargaining of transnational relations, which is a suitable complex interdependence.

Among societies, economic and social interactions will affect the groups more than security interactions. Increased transnational connections provide opportunities and are growing more costly for certain groups than others. According to some of the organisations and groups, other actors such as states and governments can benefit more than that by establishing direct interactions. The traditional approach in the context of complex mutual dependency brings to mind two questions: first, on behalf of whom? And second, what will be the result? A government bodies their interest in the name of national interest, and repeated interferences can affect the perceptions of official interest.

Finally, according to Keohane and Nye, the complex mutual dependency of multi-party connection channels is not only unique to non-governmental organisations. Governmental bureaucracies for similar purposes or simply changing perspectives, as well as certain political issues concerning non-governmental organisations, can also lead to the formation of coalitions.[51]

2.6 Compatibility with Realism and Complex Interdependence

The main assumption of the realistic approach considers the use of threat with military forces as the most effective means of international politics. At the same time, interdependence theory assumes that power of disposition determines the ability of the actor for the characteristic of each specific subject resource. Additionally, interdependence theory extended the support or hindrance of the development of interdependence as well as the use of international organisations or transnational actors for their own purposes. Otherwise, the realistic approach is that the state, which is economically and militarily strong, con-

---

50 **Keohane, Robert O. / Nye, Joseph (1977)**; Power and Interdependence, p. 11–19.
51 **Ibid**; p. 34–35.

sequently occupies a high political rank. Interdependence argues that strong states ineffectively use military force to combine different fields of action: For example, military force does not efficiently affect the price of oil.[52] In short, the interdependence approach seeks possible cooperation in order to solve international political problems.

As advocated by Keohane and Nye, the United States should play an active and leading role in the pursuit of international policy for coordination. Thus, it could be a suitable strategy for international cooperation and their stabilisation by education for international organisations, regimes and for all sides of profitable cooperative processing. This idea was developed by Keohane in the 1980s by further developing the regime theory.

For Keohane and Nye, realistic-based explanations of international politics rest on three main assumptions:

- *"States are seen as self-contained units and single most important, dominant actors in world politics.*
- *Power is the most effective means of politics: The exercise or threat of violence is the most effective method of exercising power.*
- *There is a clear hierarchy of objectives of international policy, military security (high politics) dominates over objectives in terms of economic or social issues (low politics)"*[53]

Within this idea, the realist approach is featured in world politics, with its main driving force derived from active or potential conflict among countries. In this regard, every state strives to defend its interests and borders. According to realists, this is of crucial importance for the national interest of the most powerful states. Realism argues that transnational actors are politically unimportant, and that states' main considerations are simply force or threat.

According to realist political scientists, military force is significant for all states, as this force would be supported by economic and other resources. Thus, the state would be the dominant source of power. States strive to strengthen their military in case of crises. It can be seen, therefore, that military force is a central part of national power, as Keohane and Nye emphasise in their book related to realism. On the contrary, however, Keohane and Nye also mention that the role of realism in international politics is changing. For instance, Canada considered battling the United States a century ago,

---

52 **Meyers, Reinhard (1979);** Weltpolitik in Grundbegriffen, (The World Politic in Main Components) Ein lehr und ideengeschichtlicher Grundriss, Droste Verlag, p. 304.
53 **Ibid:** 1977, p. 23.

and Britain and Germany are not seen as threats to each other anymore. In this regard, Keohane and Nye advocate that the military is not as relevant today as it was in the past, and claim that realism is not an active instrument of policy. Further, they propose that the influence of military force is costly and uncertain, although military force is continually used politically by superpowers such as the Soviets, through threats and its economic and political influence over other Eastern European countries.[54]

There are three assumptions by Keohane and Nye for the ideal type of complex interdependence:

- *"States are not self-contained units and not only actors in world politics. In addition to the States, there are other influential actors such as multinational corporations, banks or scientific expert groups.*
- *Military force is illustrated by complex interdependence at secondary option*
- *Military security is no longer a primary target in case of welfare,"*[55]

In this context, Keohane and Nye argue *"that the conditions of complex interdependence increasingly characterized world politics in some important issue areas and among some countries".*[56] On the other hand, they make complex interdependence in the approximate relationships of the western industrialised countries Organisation for Economic Cooperation and Development (OECD), especially in the problem areas of global economic and ecological interdependence.[57]

As another subject matter concerning interdependence, there is a general restriction on the relations between the western industrialised countries' distinction regarding parts of the world politics such as security, the economy and the environment, and how they play a central role. Keohane and Nye point out that the validity and applicability have two explanations, which are realism and complex interdependence. For each case there are very specific problem areas. International politics are under the conditions of complex interdependence and yet are fundamentally different from political processes identified under realistic assumptions.[58]

---

54 **Keohane, O. Robert/S. Nye, Joseph (1989);** Power and Interdependence, Harper Collins, pp. 27–29.
55 **Siegfried, Schieder/Manuela Spindler (2006);** (Hrsg.) Theorien der Internationalen Beziehungen (International Relations Theories) Opladen Hill, p. 101.
56 **Keohane, O. Robert/S. Nye, Joseph (1977);** Power and Interdependence: World Politics in Transition, Brown Company, Boston, p. 223.
57 **Ibid.;** p. 225–226.
58 **Lemke, Christiane (2001);** Internationale Beziehungen Grundkonzepte, Theorien und Problemfelder (International Relations main concepts and Theories) 2. Auflage München, p. 23.

Keohane and Nye emphasise that *"power is primarily understood as a means to regulate social relations. It is not about the unconditional consolidation of power for self-preservation of states, but is rather focused on the benefit of the parties' exertion of power within pluralistic world system. In this sense, they developed a working definition of power in the realm of Interdependence research."*[59]

In their work, power has been defined within the capabilities of the international system and extended to the analysis of non-military issues. This produces an analytical approach, which meets the increasing interdependence of states and also explains changes in the international system better than a primarily national interest on the realistic notion of power. In essence, the matter of interdependence theory shows different degrees by the interdependence of states.

Realism's main concern is military and security issues; it refuses to recognise anything but states as units of international relations. However, it became clear in the end of 20th century that military dominance was not crucial anymore. In contrast, an efficient set of states' economic policies and their ability to transform into political powers played a more central role. Interdependence suggests *"the removal of trade barriers and regional cooperation worldwide. In this case, the states find it increasingly difficult to engage in a military expansion or any kind of mannered activity because of the threat of being penalized by their trade partners. Such penalties may take the form of different restrictions such as economic sanctions such as those established against Iran in response to its nuclear program."*[60]

In spite of this focus on the demise of military dominance, the concept of power remains important in Keohane and Nye's studies. Realism claims the ultimate goal for the state is to accumulate as much power as possible so that it can provide an advantageous position in the world's balance of power; this power is reflected in the military capabilities of the country. Morgenthau assumes that *"the weight of economic resources for states in light of the changes during the last decades of the 20th century can be exemplified by the superior position in power ranking of those that are militarily weak, yet controlling essential raw material states."*[61]

According to the realistic approach, power is an end in itself, and in the world of interdependence it only serves within international relations. Keohane and Nye

---

59 **Keohane, O. Robert/S. Nye, Joseph (1977);** Power and Interdependence: World Politics in Transition, Brown Company, Boston, p. 11.
60 **Keohane, Robert O./ Nye, Joseph (2001);** Power and Interdependence, 3rd edition, Longman, p. 1.
61 **Morgenthau, H. (1974);** The New Diplomacy of Movement, in Keohane, Nye: 2001, p. 10.

define power as *"the ability to get others to do something they otherwise would not do"* or *"the actual control of the outcomes."*[62]

Nye thinks that along with military and economic power a state can have an edge in international politics because of such intangible assets such as the culture they represent, as well as their ideas and values. He also states that *"power means holding high cards in the game of international poker. But often opponent's cards are not all showing in the game of international politics. As in poker, playing skills such as bluff and deception can make a big difference."*[63] On the other hand, the state of interdependence emphasises that it is not the initial power countries possess that matters, but their aptitude to convert it into actual influence over the outcomes of interdependent relations. Additionally, when we seek to analyse international affairs, one has to take into account not only the hard military and economic power enjoyed by actors, but also the soft ones such as culture, values, institutions etc. Furthermore, Kenneth Waltz criticises mutual dependence as a myth, and states that the effect of international relations is only a marginal level of interdependence. Waltz disagrees that interdependence increases the security and peace of international relations, and believes that increasing interdependence damages the stability of the international community. Waltz asserts that among state and international communities, which differ from each other, there are arguments against mutual dependence. According to Waltz, the international structure is different from a national structure in that every state aims for the capacity of state. National structure is different from international community and contains elements such as common properties and as a result it consists of states. Moreover, the states are more or less similar to each other and therefore among states where there are no deep differences, there is no high interdependence. In addition, according to Waltz, security and military issues are affected by factors of international politics. The priority of states is to maximise their security and safety. The top priority for states is survival and therefore for states, nothing can replace this priority. However, low policy, which focuses on prosperity and welfare, is at a secondary level compared to security on high policy. According to Waltz, it is not possible to connect two different areas.[64]

---

62 **Keohane, Robert O., Nye, Joseph S. (2001);** Power and Interdependence – 3rd edition, Longman, p. 10.
63 **Nye, Joseph (2005);** Soft Power: The Means to Success in World Politics, public Affair Pr., p. 59.
64 **Morse, Edward (1972);** "Transnational Economic Processes", Robert O. Keohane and Joseph S. Nye, Transnational Relations and World Politics, Massachusetts, Harvard University Press, p. 30–33.

3. Energy Interdependence as a Challenge

When we consider Turkey and the EU's gas dependence on Russia and other states, we can identify that importers can incur costs such as supply disruptions and can also increase costs for exporters in the short term. For example, exporters can expand the gas storage capacity. The dependence between importers and exporters is mutual, and by their actions, both sides are able to cause the other costs. This can happen when the price changes on the related or delivered gas volumes, or because of the special characteristics of the gas trade, such as pipeline boundaries, long-term contracts and oil prices.

The importer cannot reduce the amount of gas without authorisation, and cannot switch suppliers or dictate unreasonably high prices. Any breach of the contractual framework, whether intentionally or unintentionally, is associated with high costs. There are high costs in the long term; such as the credibility and reliability of a partner and the inevitably of other losses and damages

Mutual dependence has influenced entities to adhere to long-term contracts and has clearly contributed to the shape of future relations. The exporter can decide before a supply agreement and related infrastructure has been built. The potential importer may also decide whose natural gas to import, what price to pay, and also via which route the natural gas will flow. In this context, one should recall that natural gas is used as a substitute for petroleum products or coal. This also means that natural gas is a candidate for substitution even when the other sources of energy, for reasons such as economics, technical supplies, environmental considerations, etc. seem more attractive.

In bilateral gas relationships, there is only one importer and one exporter, and the exporter holds a better position than the potential importer before the conclusion of contracts for the exporter. An example of a bilateral gas relationship is the Azerbaijani decision for Shan Deniz phase I–II, where long-term contracts were signed and then the pipeline was built. An advantage such as this lies with the importer.

The medium- and long-term framework for gas export or import in the future leads to high costs for the players, such as consumption of producers, spending gap and diversification, as seen with Turkey. The price can reduce according to the energy diversity, renewable energy or energy efficiency.

In situations such as these, the actor or group of actors and new rules can be changed. New rules apply only to finished contracts and shall be decided by joint negotiation. In the past, this practice was different, particularly since there is a benefit for stakeholders such as importing and exporting corporations. However, players have moved outside of these rules, as seen with Algeria during the gas fiasco and Russia and Ukraine in past gas disputes. This connection was made from one side in the attempt to exploit a position of relative strength by bending rules or by actors moving beyond existing rules. This type of approach is rare because although the actors know that they are able to draw short-term benefits at the level of interdependence sensitivity by methods such as blackmailing for higher prices, they also take into account the high costs associated with these methods, for example in the case of interdependence vulnerability, because their partner's natural gas needs can be met by other sources.

Concrete interdependence relationships between the EU and importer countries should be considered advantageous, an example of which are the relationships of the individual actors in gas sector. Furthermore, their interests and goals should be analysed and evaluated based on how the implementation of these goals and interests affects the goals and interests of the other. The costs in particular must be analysed in order to determine how each actor could incur potential costs to the other. As transit states are significant actors, their role must also be investigated, and placed in the context of relations between the EU and Turkey and the relevant exporter or the respective exporting countries. On the one hand, the interests of the individual transit states are considered; on the other, the interest of the EU and Turkey and those of the exporting country must be assessed.[65]

3.1 Pipeline Transit State on View of Vulnerability

To be more precise, an analysis using the term 'vulnerability' cannot be restricted only to relations between the European Union and Russia. Pipeline transit states play a central role in European energy policy on natural gas. In particular, the planned pipeline projects such as the TANAP and TAP projects generate new interdependencies between the cooperating actors. If we consider the Ukraine–Belarus gas crisis with Russia, which increased gas prices, and Ukraine's exposition, then a sense of political sensitivity would develop. This would be sensitive for both sides, but especially for Ukraine.

However, if Russia had not been willing to comply with the wishes of Ukraine and threatened Ukraine, Ukraine would have been highly vulnerable and reached the point where continues energy policy. The East–West conflict in the second half of the century has set new standards of interdependence. The areas of energy

---

65 **Keohane, O. Robert/S. Nye, Joseph (1977);** Power and Interdependence: World Politics in Transition, Brown Company, Boston p. 12–19.

policy, oil and gas in particular are being used as a new weapon. With the 1973 oil crisis, oil and gas were defined as a modern weapon. It raises the question of the characteristics and the structure of interdependence in natural gas supply to Europe. This crisis should be mentioned and analysed, particularly in the gas sector, at a sensitivity and vulnerability level. *"Power and Interdependence"*, written by Keohane and Nye, takes historical changes of international regimes and explains these changes using different approaches from the interdependence structure. This current situation is assessed in relation to the security of European gas. If we inspect this situation with interdependence in terms of the energy relations in the gas sector, keeping vulnerability and sensitivity in mind, it would be useful to attempt to take a look at possible future developments. The focus of interest in this context is primarily the creation of a gas cartel and close cooperation between gas producers and exporters with cartel consequences for Europe.

Russian and European partners avoided transit countries like Ukraine and Russia, instead they developed and planned the South Stream and Nord Stream pipelines, which will effectively reduce the weight of the Ukraine as a bottleneck in the gas transit. Two pipelines to the participating actors are the answer to the vulnerability; however, Nye and Keohane ponder alternatives and their costs. The major political changes in Ukraine in recent years, in addition to the associated gas crisis with Russia required a policy change. The pipeline route planning for future projects between Russia and the countries of the European Union today is designed to not pass through transit countries. The cost of this policy adjustment is high. The pipeline at the bottom of the Black Sea the South Stream and the Baltic Sea Nord Stream are significantly more expensive than a pipeline along existing routes.[66] Within this context, an alternative policy has been developed as Keohane and Nye suggested: The TANAP-TAP pipeline is an alternative to Nabucco starting at the Georgian–Turkish border. The Shah Deniz Consortium decided to join the TAP pipeline project in the final phase. It will bring Caspian gas from the Turkish--Greek border to European recipients.

In conclusion, the EU and Turkey are looking to decrease their level of vulnerability and dependence on Russian gas. The EU seeks to improve the security and diversification of their supplies in the Southern Gas Corridor. Not only is the energy demand for gas increasing, but production of alternative energy resources is expensive and it is not seen as an easy task for the EU and Turkey to diversify their suppli-

---

66 **Katharina, Graisy (2009)**; Die Interdependenz zwischen Russland und der Ukraine unter Berücksichtigung der Ressource Erdgas, (Interdependence between Russia and Ukraine under Considering of Natural Gas) Vienna, p. 17.

ers. When we take these factors into consideration, the situation could be explained as vulnerability interdependence, which leads to costs for the EU and Turkey.

3.2 Role of the Actors in Gas Policy

Today, there are usually economic and military interests among actors that lead to interdependence. In this case, the actors are not enemies, but competitors who benefit from each other. According to the traditional state-centered approach, technology, geography level and domestic policy ensures relationships among states. However, the transnationalist approach points out that inter-state relations not only concentrate on the traditional approach. Instead, relations among states and transnationalism implies commercial relations, communications, tourist and business trips. Thereby, if we consider relations among states in international relations in a framework of this aspect, states are not the only actors of international relations. There are also influences beyond state institutions on international affairs, such as religious communities, oil companies and non-state institutions. Transnational relations consider non-state institutions as influential in international political process, and also point out that they are in competition with many issues among states; therefore, non-state institutions are seen as actors according to transnationalism. The transnational relations paradigm of international relations not only sees military and security issues on the agenda, it also examines significant economic and social relations with regard to transnationalism. Within this framework and on the basis of mutual dependency, there are relations among actors in open and hidden bargaining processes.[67]

Most or some wars, such as the commercial and financial relations initiated by the governments of sovereign nation states and their control of the global interaction by developing and classic diplomatic activities, carried out interstate interactions. These interactions do not simply limit interstate or non-state institutions and organisations; they involve governments and are seen as transnational executive interaction, and non-governmental actors play a prominent role in this process. It is not important whether the actors are governmental or non-governmental, since the resut is that the interaction of international relations goes beyond borders. Therefore, it is called transnational interaction and communication.[68] According to Nye and Keohane, it is difficult to determine the

---

67 **Torbjorn, L. Knutsen (1992);** History of International Relations Theory, Manchester University Press, p. 235–36.
68 **Nye, Jr., Joseph S./Robert O. Keohane (1972);** Transnational Relations and World Politics: An Introduction, Cambridge, Massachusetts Harvard University press, xii.

role of governmental, intergovernmental or non-governmental actors, and where the interactions begin and end. It can be difficult to predict whether an actor acts as an official employee of government or on its own behalf. Moreover, many non-governmental and local organisations, which also operate activities in other countries, can be considered as transnational actors, such as journalists, students, farmers and national trade unions. Thereby, cross bordering of all the activities of these organisations is not necessarily implied. The operation of Unilever in the United States, IBM in Brazil and Standard Oil Company in New Jersey (Exxon) established a formal relationship with the French government, which, of course, should be evaluated within the scope of transnational relations.[69]

In this sense, as executing agents operate privately, partly public and public energy companies are trying in their own interest to improve their economic performance and increase their revenue. From this motivation at the economic level, the TANAP and South Stream were planned, and energy companies have joined together across borders. For example, SOCAR (state-oil company of Azerbaijan) and BOTAS (state-owned crude oil and natural gas transportation and trading company of Turkey), while the South Stream Pipeline, which was initiated by Gazprom and the Italian oil company ENI from the Italian side, has already been negotiated at the highest political level and support. TANAP, which was signed in 2012 by the states concerned, was also formulated. Nabucco, after many years of uncertainty obtained the necessary legal and political guarantees, but the difficulties for Nabucco could not be overcome. Thereby, Azerbaijan and Turkey have begun an alternative project, which aims to start in 2013. Here, public energy companies are seen as significant actors, for example, BOTAS and SOCAR are backed by the state as Keohane and Nye suggested.

As a conclusion, the energy companies which have a role to play in the implementation of the projects also have great influence within the political and economic policy in the process. In such pipeline projects, the multinational energy companies that execute the projects are as effective as the states themselves. The lobbying actions carried out by these companies have a direct impact on these decisions. It should not be ignored that some companies who are present in the Shah Deniz Group are also partners in the TAP project, which had a great impact in the selection of the TAP project. Besides, it is planned that other members of the Shah Deniz Group will also participate as partners.

---

69 **Ibid,**: xvi.

3.3 Diversification of Natural Gas Supply for the EU and Turkey

The annual rise of Turkey's electrical energy demand is at the rate of 7–8%. The ministry of energy and natural resources claims that external energy dependence will increase in direct proportion to the population growth and it is a necessity to construct a nuclear energy power plant.[70]

In Turkey, consumption of natural gas has become increasingly widespread since 1987. In total primary energy supply, the share of natural gas reached 32% in 2011. It is estimated that the natural gas usage level will reach 82.7 billion cubic meters in 2020. Currently, Turkey imports 57.9% of its natural gas demand from Russia, 18.7% from Iran, and 8.7% from Azerbaijan. In addition, Turkey has several liquefied natural gas (LNG) import agreements with Nigeria and Algeria. Forty-eight percent of imported natural gas is used in the electric power production. When examining the sectoral allocation of national natural gas consumption in 2012, the breakdown is as follows: electricity 48%; industry 26%; and housing 25%.[71] Petroleum is imported from Middle Eastern countries, notably Iran and Saudi Arabia in addition to Russia. Finally, Turkey's demand for energy has risen at a rate of 8% annually. With this growth rate, Turkey possesses one of the biggest growth rates around the world. Turkey has to generate some solutions in response to this rising energy demand.[72]

Similarly, according to EU Eurostat data, whereas the EU's gas import was 16% in 1999, it reached 64.2% in 2009. In case that this increase continues, it is estimated that the EU will have 80% foreign gas dependency in 2030. In addition to this, the reserves in the North Sea will gradually decrease while the gas consumption increases. Whereas the EU was dependent on Russia for gas by 34% in 2009, Norway and Algeria are the main suppliers of the EU with 31% and 14% respectively.

In this sense, when we look at the relationship between the Soviet Union and the West since the foundation of the Soviet–Western European gas trade, interdependence could be defined as a classical sort of dependence. By this definition, *"each side held a degree of power over the other. At the same time, the gas was provided by the Soviet Union and the Europeans were the source of hard-currency*

---

70 **ETKB** – Documents about Nuclear Reactor Planning of Turkey (Republic of Turkey Ministry of Energy and Natural Resources), p. 5.
71 World Energy Council Turkish National Committee: Energy Report Ankara, 2010, p. 53.
72 **EPDK (2012)**; Turkish Energy Market: An Investor's Guide, 2012, p. 31.

payment and equipment deliveries."[73] In their work, Robert Keohane and Joseph Nye argue that *"[i]nterdependence is a useful theory when analyzing the EU-Turkey transit status on energy security. As they point out, interdependence is a mutual dependence, and in world politics this refers to situations with reciprocal effects amongst countries or actors in different countries."*[74] These interactions are dependent on the type and the strategic importance of the raw materials being traded.[75] It is known that Russia uses energy as a threat against the EU, which is why the EU is looking for alternative routes and energy sources. At this point, *"Turkey's further development as an energy bridge and potential energy hub,"*[76] as noted by the European Commission, *"will benefit both Turkey and the EU."* A further market integration would be of mutual interest for both sides. The EU's gas supply and energy needs would be met more by the imports in the future. This situation makes the relationship between the EU and Russia more complicated. One of the first important issues on the political agenda in the EU has become the security of its energy supply. In this regard, it can be analysed that the high sensitivity of the EU's gas dependence on Russia, in effect, decreases the level of vulnerability. On the other hand, the EU's vulnerability with regard to a potentially impending gas crisis will have a destructive impact on the EU's economy and industry. This risk is why it is imperative that the EU seeks other gas supplies from North Africa, the Middle East and the Caspian region to diversify its energy resources. In this respect, Turkey holds a very strategic position to secure and diversify gas and oil supplies for the EU.

Russia has a history of using its gas supply to its advantage: the Ukrainian Crisis in 2006 was caused by the use of natural gas as a Russian foreign policy tool in its relations with the Ukraine. It is common knowledge that the Ukraine is the gateway for nearly 80% of Russian gas exports to Europe. This crisis harmed Russia`s reputation as a reliable supplier and placed the Ukraine in the position of having insufficient supplies of natural gas to maintain its own gas needs. The problem of Russia's gas pricing resulted in the cutting of the supply of gas to the Ukraine. The lingering question is, why did Russia behave this way towards the Ukraine? Furthermore, did it risk its reputation within the EU? The term 'asymmetries of

---

73 **Jonathan, Stern (1987);** Soviet Oil and Gas Exports to the West: commercial transaction or security threat?, Gower Pub. Co. Energy Policy Studies, p. 59.
74 **Robert O. Keohane/Joseph, S. Nye (2001);** Power and Interdependence: World Politics in Transition, p. 7.
75 **Ibid.;** p. 8.
76 **European Commission:** Turkey-EU Positive Agenda Enhanced EU-Turkey Energy Cooperation, outcome of meetings- 2012.

dependence' can give insight to these answers. As put forward by Keohane and Nye, asymmetries of dependence *"is most likely to provide sources of influence for actors in their dealing with one another."*[77]

On this point, the EU and Turkey in the gas sector are in a close interdependence. The planned pipeline Shah Deniz II has expanded to the border of the EU with TANAP-TAP. Both countries can benefit significantly from this pipeline, as it is the only one among the major suppliers of gas to the EU within this strategy of diversification. Azerbaijan gas exports via pipeline with the transit countries need to go to Europe. Kadir Dikbas, the economy columnist in Zaman newspaper, said in this regard that *"Turkey is affected by both the crisis in Europe and the political development going on in the Middle East. Besides, it has suffered from the rise in prices of the petroleum as a power importer."* The columnist also indicated that *"…the most important items in energy import are the petroleum, natural gas, fuel and diesel oil. The import of these items in 2009 was around 29.9 billion dollars but has risen to 38.5 billion dollars in 2010. Finally, Turkey's energy import has risen to 42.2 billion dollars in 2011 according to the foreign commerce data."* As a summary, the columnist emphasised that *"Turkey's total merchandise trade deficit is over 90 billion dollars and 44 billion dollars of this merchandise trade deficit is paid for energy import".*[78]

In summary, Turkey's strategic position and dynamics can contribute to the diversification of supply for the EU, in addition to the diversification of transit routes for Russia. It is demonstrated that an interdependent relationship between the EU and Turkey can have some costly effects on TANAP in the future. Therefore, this interdependence should be constructed in a way that is satisfactory for both parties.

3.4 Cooperating for Energy Security Between the EU and Turkey: An Interdependence Approach

This dissertation considers the relationship between the EU and Russia in the context of gas imports, and the relationship between Turkey and the EU as part of the transit status of Turkey. The theory of interdependency may also be explained under the title of 'security theories' because of their interrelatedness. The EU suffers a lot from its dependency on Russia. Today, it is the most significant external supplier of gas to Europe and Turkey. Conversely, Russia is dependent on the EU and Turkey, as Europe and Turkey are some of Russia's biggest economic partners.

---

77 **Keohane/Nye (2001)**; p. 9.
78 Kadir Dikbas: http://zaman.com.tr/yazar.do?yazino=1209175 (accessed on 28/12/2012).

In terms of a Waltzian notion of security, *"although interdependence is usually seen as 'sensitivity'"*, he rejects it and offers a definition of interdependence as *"mutual vulnerability"*.[79] Within this context, he also added that defining interdependence as sensitivity leads to an economic interpretation of the world. In this regard, understanding the foreign policy implications of interdependency necessitates concentration on the politics of international economics, not on the *"economics of international politics"*. Moreover, interdependence means that the parties are mutually dependent; thereby there is reciprocity among them, rather than a one-sided vulnerability. As Waltz puts forward, with reference to the 1973–74 oil crisis, the oil importer countries went on a major strike, which resulted in a crisis revealing their dependency on the oil and gas exports. In effect, *"the political clout of nations correlates closely with their economic power and their military might."*[80]

In this crisis, the western European countries had to consider economic necessities, whereas the USA could manipulate the crisis so as to foster a balance of interests and force holding a prospect for peace; the US's actions in this crisis show that some nations have little ability to affect events outside their borders, while some have an enormous ability to do so. Yet, the interdependency in the globalisation era is more of a reciprocal issue in the sense that no single country, even a superpower, can dominate others without any repercussions. Just as Europe needs Russia for gas supplies, Russia needs Europe to purchase its gas. Likewise, as Turkey needs Europe for the membership benefits, the EU also needs Turkey, for nothing if not to secure the energy routes. As former Commission vice-president Verheugen said, *"Europe cannot do it without Turkey and The European Union needs Turkey if it is to succeed as a global player."*[81]

Interdependent countries need each other in order to function, and both sides of the equation hold significant benefits for the other; otherwise, the relationship would not be interdependent, and would even questionably exist. For this reason, Keohane and Nye made further interpretation of interdependence and define it as mutual dependence.[82]

As a matter of fact, they developed a concept called *"Complex Interdependence"*, which is also relevant to energy geopolitics. Unlike realism, which attributes high importance to military security as the dominant goal, and military force as the

---

79 **Waltz, Kenneth (1979);** Theory of International Politics, Addison-Wesley Pub. Co. p. 139–146.
80 **Ibid.;** p. 139–146.
81 http://www.euractiv.com/future-eu/verheugen-turkey-eu-reforms-esse-interview-518777 (accessed on 10/03/2013).
82 **Keohane/Nye (2001);** p. 7.

most effective instrument, complex interdependence highlights that the aim of states may vary by issue areas, and the power resources specific to issue areas are more relevant than military resources.[83]

In complex interdependence, state policy goals are not according to stable hierarchies but they depend on trade offs. *"The existence of multiple channels of interaction among societies increases the range of policy tools and limits government's control over foreign affairs; moreover, in this point of view, military force is irrelevant to a great extent."*[84] In the case of Russia, the officials are aware that Russia is no longer a military superpower, hence if it wants to accumulate power it has to "play" another card, namely being an energy superpower and using its vast oil and natural gas resources. In light of this, the nature of the EU's energy relations with Russia as interdependent was described by the previous Energy Commissioner of the European Commission, Andris Piebalgs, as follows: *"The relationship is one of interdependence not dependence, which means that Russia needs us as much as we need Russia."*[85]

Hence, energy security can best be understood in terms of an approach to embrace the interdependency between the parties at issue. It is nearly impossible to come to any agreement without a decision-making dialogue amongst all states. This is, of course, a result of globalisation. In this regard, countries and continents developed mutual dependence within the energy pipeline projects, and the integrity of politics and economics affects domestic policy decisions. This means that the effects of domestic politics can bring immense damage to the foreign policy of states with themselves, with domestic policy at the micro level and foreign policy at the macro level.

The analysis of the context of EU-Turkey energy relations explains that outcomes of such interaction can be defined as cooperative. Within an interdependence theory framework, the reasons why cooperation is needed, as well as why cooperation occurs, can be seen in the example of EU-Turkey relations. Both the EU and Turkey realise the significance of integration in the field of energy for the enhancement of energy security. Cooperation led to positive results and achievements like the reduction of pollution, trade of nuclear materials, creation of energy efficiency programmes and the Kyoto Protocol by Turkey, as well as the initiation of several projects for companies such as trans-European energy networks. Both

---

83 Ibid,; p. 32.
84 Ibid,; p. 32.
85 EU-Russia Center, "The EU-Russia Center Review: EU Russia Energy Relations", Issue 9, June 2009, retrieved from http://www.eurussiacentre.org/wpcontent/uploads/2008/10/review_ix.pdf (accessed on 05/0472013).

parties have defined rules and principles. The European Commission's multilateral cooperation within the energy field began the initiation with Turkey. Turkey decided to implement the European Commission Treaty in order to evaluate all possible outcomes of the ratification development in a legal framework. In terms of interdependence theory, in this case, Keohane and Nye discuss the relationship between states and multinational corporations or unilateral agreements between countries such as TANAP and the proposed existing pipeline TAP. They combine unilateral agreement so that it can be defined in the Southern Corridor with interdependence theory. In this context, Keohane and Nye point out that *"[s]tates are not self-contained units, and not sole actors in world politics"*. In addition to the states, there are other influential actors such as multinational corporations, banks or scientific expert groups that provide *"multiple channels of contact."*[86] States and their political entities are accompanied by power plays and power processes, and these mutual relations can be explained with an interdependence approach, as the theory is related to domestic and foreign political events. The ever-increasing global trade relations and division of labour are contributing to the need to adapt to these new circumstances. The result of these collaborative relations is reciprocal relationships, which bring mutual cooperation and less dissent with each other, which in effect causes fewer conflicts at an intergovernmental level.[87]

In the context of energy dependence, the diversification of gas is significant with regard to interdependence. The EU imports about 80% of their natural gas through Ukraine, creating a one-sided dependence. This situation cannot be explained with the interdependence theory because it is one-sided and therefore not interdependent. In this point, the EU and Turkey can diversify their natural gas needs via the Southern Gas Corridor, which reflects interdependence because both sides need each other in term of diversity.

3.5 Contributing Energy Cooperation of Turkey to EU Membership

In December 1964 Turkey became an associate member of the European Economic Community (EEC) with the prospect of full membership. After 40 years, at the December 2004 summit, the EU declared that Turkey had fulfilled the Copenhagen Political Criteria and could start accession negotiations. Relations

---

86 **Robert O. Keohane/Joseph S. Nye (2001);** Power and Interdependence: World Politics in Transition, Longmann pp. 23–24.
87 **Lehmkuhl, Dr. Ursula (2001);** Lehr und Handbücher der Politikwissenschaft: Lehrmkuhl, Theorien Internationaler Politik, Oldenburg, 3. Auflage, p. 194.

between Turkey and the EU changed after Germany and France's statements about a privileged partnership instead of a full membership with Turkey.

The critique on the EU enlargement, by means of rationalist and constructivist theories, questions why the member states agreed to the EU enlargement, although negative effects and high economic, institutional as well as security policy costs were linked with it. States are the most important players on the international level. This is from where the liberal, most relevant and rationalist approach for the clarification of specific decisions in the EU emanates. State behaviour and acting is determined by the logic of usage maximisation. Costs and usages of economic interdependence are considered as important factors in influencing state preferences. Governments define their interests and preferences in their own country, in order to realise them later on at an international level in negotiations between the governments.[88] In this way, Europeanisation emphasises the internalisation of EU norms, direct or indirect pressures from the EU both in member and candidate countries and constructivists draw attention to society rather than nation states. Constructivism takes the Europeanisation as a process of construction into account, and acknowledges that cultural dynamics play an important role. Additionally, constructivism suggests that it is the normative arguments rather than rational arguments because of the critical decisions that play a substantial role regarding the future of the EU, like eastern enlargement. Consequently, this thought also highlights the role of identity in decision-making processes within the EU, indicating the different responses of Germany and the UK to the adaptation of a single European currency.[89] In this respect, this thought refers to the influence of globalisation and takes into account the different logics of the construction of Europe and also the importance of norms and identities in decision-making.

Featherstone and Kazamias state that the concept of Europeanisation is divided into three dimensions. The first dimension is the increasing and expanding of institutionalisation in the EU. The next dimension is the incorporation of norms, rule identities and interests of actors at the level of member states.[90] The last dimension is the adaptation of EU norms by non-member states and it also includes candidate states, like Turkey.

---

88 **Moravcsik, Andrew (1993);** Preferences and Power in the European Community: A Liberal Intergovernmentalist Approach; Journal of Common Market Studies, pp. 480–482.
89 **Özcan, Mesut (2008);** Harmonizing foreign policy, Turkey, the EU and the Middle East, Ashgate, Aldershot pp. 21–22.
90 **Featherstone, Kevin/ Kazamias, George (Ed.) (2011);** Europeanization and Southern Periphery, Frank Cass, London, pp. 5–6.

The EU and candidate countries make an effort to compulsorily or coercively influence the candidate countries to adopt EU norms as their domestic structures. Secondly, the candidate countries are forced to comply with the conditions of the EU as a result of the uncertain outcome of the accession process. Finally, asymmetrical relationships allow the EU to determine the rules of the game in the accession negotiations and therefore there is no bottom-up process of Europeanisation in candidate countries.[91]

In this point, Turkey must adopt European legislation, including those in relation to the energy markets. Thus, the EU has an important impact on the energy sector, which is of economic importance for Turkey, especially with regard to Turkey's candidacy to become a member state of the EU and its participation in European and other international energy companies. In addition to the transparency of the sector and the participation process, a high level of security ensures that the interdependence relationship in the gas sector between the EU and Turkey is as good as symmetrical. Both sides benefit from working to a high degree. New projects have supported the position of Turkey, especially as this collaboration has been made in the context of clear and established rules such as those of the European Economic Area. Furthermore, there is little opportunity or reason to exploit interdependence sensitivity or even vulnerability. The potential cost on both sides would be, at any level, very high.

Turkey plays a key role for the EU in a route to access Caspian, Middle Eastern, and other Southern and Eastern resources. Another point that *"contributes to the significance of Turkey is its accession process to the EU. Turkish energy legislation adapted the laws of the EU internal energy market on 18 April 2001 and ratified the Energy Charter Treaty in 2001. Confirming Turkey's strategic position, the European Commission pointed out the importance of Turkey in the transportation of Caspian and Middle Eastern gas to Europe".*[92]

As Güentiger Oettinger, European Commissioner for Energy, expressed at the meeting of the Konrad Adenauer Foundation in Brussels: *"I wager that in 10 or 20 years' time, the German Chancellor and the President of France will go down on their knees begging Turkey to join the EU. 'Friends, come to us'".*[93] He mentioned in

---

91 **Heather, Grabbe;** Europeanization Goes East: Power and Uncertainty in the EU Accession Process, in; **Featherstone, Kevin/ Radaelli, Claudio M. (Ed.): (2003);** The Politics of Europeanization, Oxford University Press, Oxford, p. 303.
92 **European Commission, Green Paper (2001);** "Towards a European strategy for the security of energy supply".
93 http://www.bild.de/politik/ausland/guenther-oettinger/eu-kommissar-kritisiert-eu-kurs-zur-tuerkei-29190992.bild.html, (accessed on 14/04/2013).

his speech that economics are an important part of Turkey's accession to the EU. Turkey's membership may create a win-win situation for both sides, and political gains are also possible. In this case, Turkey's EU membership would utilise Turkey's influence in a wider region, from the Balkans to Central Asia and the Middle East. In this sense, energy security is one of the key dimensions of Turkish membership in the EU.[94]

Another agenda of accession negotiations is the integration of the Turkish gas market to the EU. Turkey's aim is that its contribution from energy cooperation with the EU will aid the negotiations. Furthermore, the EU itself links the pipeline to interests other than economic benefit. The EU Commission appears to embrace the link of energy cooperation and membership, and as President Barosso stressed, *"Turkey can in fact be something that is in the interest of all European citizens: Good cooperation on energy matters."*[95]

Further interconnection between the EU and Turkey would be of mutual benefit to both sides in energy matters. In this sense, the opening of an energy chapter to negotiations could provide important momentum in terms of further alignment of the EU's internal gas market with Turkey. On this matter, the Commissioner for Enlargement Olli Rehn stated, *"the EU and Turkey share essential strategic interests, e.g., in security, economy and dialogue of civilizations. That is one of the reasons why the EU decided to open negotiations for membership with Turkey. Energy strategy is an area where both the EU and Turkey can gain from deeper cooperation."*[96] In the Black Sea Power and Economy Forum, Prime Minister Erdogan criticised the EU saying that *"Turkey will be an energy corridor and it will be one of the most important power centers of the world in near future. Furthermore, the EU has still not opened the Energy chapter although Turkey and the EU are in a negotiation period and the EU uses this situation as a threat to Turkey. However, not only Turkey but also the European Union will get the best of Turkey's full membership in the EU. This wills a mutual advantage for both sides."*[97]

---

94 **Julian, Horn-Smith (2008);** "Turkey: Trade and EU Accession", **Adam Hug** (ed.), Turkey in Europe: The Economic Case for Turkish Membership of the European Union, London, The Foreign Policy Center, p. 51.
95 **Energy pushes Turkey and EU closer:** http://www.nytimes.com/2009/01/19/world/europe/19iht-turkey.4.19499151.htm, (accessed on 20/01/2009).
96 **Joint Press Release, Turkey and the EU (2007);** Together for a European Energy Policy—High Level Conference.
97 http://www.haberler.com/basbakan-erdogan-enerji-dagiliminda-adalet-2258967-haberi/ (accessed on 13/02/2012).

Finally, if economic ties are strengthened by such projects, nations will be able to connect their populations and strengthen political interaction among nations. These economic relationships among nations can be explained with the interdependence approach. This focuses on welfare, opening market access and increasing trade. To sum up, economic interdependence affects the political behaviour of states. In this point, the argument of this dissertation is that further cooperation on common projects will affect the political behaviour of states. Thus, Turkey's membership in the EU will accelerate through these projects.

It is argued that the EU and Turkey want to concentrate more on their economic, mutually beneficial cooperation, and primarily care about the welfare of their respective citizenries. They will perhaps construct a common front to tackle global challenges such as the environment, terrorism, food and financial crises, but using a military concept is unthinkable given the destructive potential of both sides. As guider of liberalism Immanuel Kant particularly emphasised, international anarchy can be overcome by collective action like a federation of states and regional blocs etc. Kant considers that through this cooperation, humans will learn to avoid war; however, it will not be easy. Liberalists hold an important place in international institutions such as the League of Nations and the European Union. According to liberalist assumption, such institutions can solve problems and provide collective cooperation in a multilateral form. Moreover, if viewed independently from a broader picture, EU–Turkey relations do follow the pattern and uphold the theory of interdependence, mutual dependency and ample interconnectedness. Furthermore, the relations are within the global context of changing international politics and also lack of global governance, causing energy to emerge as a hard political power.

Summary

In this study, the mutual dependence relations are explained with the theory of interdependence taking into account the transit role of Turkey within the framework of EU and Turkish gas import from Russia. Within this framework, the theory of interdependence aims at increasing cooperation between interstate or interaction actors and mitigating the conflict environment. Keohane and Nye, who are the theoretical defenders of the theory of interdependence, claim that in the realist theory, Hans Morgenthau ignores the functional relationship between political, military and economic powers. Keohane and Nye clearly demonstrate the relationship between power and interdependence, indicating that the parties need cooperation in the relations between them.

Accordingly, under the interdependency conditions that arise in the international system, the bargaining power of one of the parties on the other is linked with the sensitivity and vulnerability of this party towards the interdependence relation.

Whereas the sensitivities of the USA, Japan and European states towards Middle East oil and of the EU and Turkey towards Russian gas are more or less at the same degree, their degrees of vulnerability against any possible negative developments in relation to the resources in the country are not the same.

As trade relations gradually develop with globalisation, interdependency also increases continuously. These relations, on the other hand, point out the development of neoliberal free trade. The developing relations also increase the cooperation, mitigating the conflict environment. Neofunctionalism and functionalism theories have parallel ideas with the theory of interdependence. The shortcomings experienced between Turkey and Iran and the crises that have arisen from time to time between the EU and Russia lead the governments to change their foreign policies. Even more, the increasing strategic importance of Turkey and its boosting gas consumption accelerated the accession process in EU-Turkey membership relations.

In this context, the opening of the Southern Gas Corridor with the developments brought by the TAP project following the TANAP project initiative will be continued by other projects. As it is put by the theory of interdependence, states are not single isolated actors in global politics; other actors also have a role to play in politics (such as companies, banks etc.). This interdependency and the commercial relations between the governments also bring about political union. International projects lead to change in the foreign policies of the countries. The importance of public participation companies in the Southern Gas Corridor cannot be ignored. At this point, we can count companies such as Gazprom, SOCAR and BOTAS as significant actors with the states.

As a conclusion, Turkey could control energy in its international relations as an important tool in political, economic and strategic terms as a bridge of energy, and thus bring solutions to international problems.

4. Global Growing Energy Demand

4.1 Increasing Importance of Gas

This section will explain the growing use of gas in the world, and appropriate ways for gas to be used in the future. While the modern world community works largely through the anarchic, self-help, nation state system, there are many international and transnational associations to facilitate cooperation between its members. As Seyom Brown emphasises, some of the specific and regional associations are highly institutionalised, namely formal assemblies and permanent secretariats. As world natural gas consumption is growing nearly 2% yearly in the world, the OPEC and such institutions are continually rising in importance at this time because of any disruption or damage that the gas situation can cause to exporting countries. Furthermore, Brown states that almost every country is a member of the United Nations, either directly or indirectly: there are different world organisations for peace, security, economic development, health and environmental protection.[98] If we consider cooperation as a behaviour between two or more actors, then we will see a wide range of international cooperation among national governments or other actors across national borders; cooperation which has been taking place under the anarchic structure of a nation state system.

Twenty percent of the world's population is composed of developed countries. While these countries strive to maintain their level of development, the remaining 70%, that is developing countries, aim to increase their level of development in the first half of the 21st century. Coal with 24–26%, liquid oil with 40% and gas with 22–26% were ranked as the main energy resources,[99] demonstrating that the energy sector is one of the most significant and indispensable necessities of humankind. The welfare level of countries is determined according to not only per capita income but also per capita energy consumption. Additionally, the energy sector is considered to be a developmental tool.[100]

---

98 **Brown, Seyom (1996);** International Relations in a Changing Global System: Toward a Theory of the World Polity, Westview Press, p. 28.
99 **Özer, Serra (2004);** A Feasibility Study and Evaluation of Financing Models for Wind Energy Projects: A Case Study on Izmir Institute, A Dissertation, Izmir Institute of Technology, p. 1.
100 **Demirbas, Ayhan (2003);** Fuel Wood Characteristics of Olive Huskand Walout, Sunflower and Almound Shells, Energy Sources, No 25 p. 310.

As it is emphasised in the TUBITAK vision report of 2023, it is foreseen that fossil fuel dependency will increase towards the midcentury, and, due to the concern about depletion of these resources, competition over oil and gas will become drastically more competitive. All countries know that the security of supply can be provided if the required energy resources are supplied continually at affordable prices.[101]

The constantly increasing energy demand in developing and growing economies requires the countries to diversify their resources and use them in an efficient way. Today, coal, oil, gas and nuclear energy are the first ones to come to mind when talking about the main energy resources. Energy resources such as oil and gas have relatively important roles among other resources, since they are only found in certain parts of the world and possess more value. Firstly, by 1973, oil, which is one of the most prominent energy resources, had gained increasing importance. Then nuclear energy became popular and recently gas has taken its place among the most important energy resources.

In a symposium report published on this subject, Mustafa Kibaroglu points out that the states are either being more protective over their energy resources, or are striving to have control over the regions with energy reserves so that they can have an active role in the production and distribution of these resources. Kibaroglu also thinks that many states determine their political strategy based on this purpose. For instance, according to him, one of the underlying reasons for the US occupation in Iraq is the desire to get those regions with immense energy resources under their control. Likewise, each state aims to be self-sufficient in terms of energy resources and wants to supply energy without any dependency on other countries. The widespread belief is that the less a state is dependent on other countries in terms of energy resources, production and supply, the more self-confident and secure it is.[102]

States aspire to improve the economic and political conditions of the country by striving to be independent in terms of energy supply. While oil and gas reserves abound in the Middle East, the population and the amount of per capita energy consumption is conversely low. On the other hand, Japan, which is a country with a huge population and high amount of per capita energy consumption, does not have oil or gas reserves. To sum up, in the regions that own rich energy resources,

---

101 TÜBITAK; Vision 2023 Technology foresight project energy and natural resources report p. 1.
102 **Kıbaroğlu, Mustafa (2004)**; Dünya ve Türkiye deki Enerji ve Su Kaynaklarının Ulusal ve Uluslararası Güvenliğe Etkileri, (Energy and Water Resources in the the World and Turkey and their Impacts on National and International Security).

per capita energy consumption is much lower due to the small population and the level of development. In this sense, there is a non-proportional distribution of energy resources in question. The basic reason why countries war with each other is the desire to get the energy resources under control and to be in possession of them. Recently, especially with the industrialisation of developing countries such as China, India, and Latin America and with the increase in the energy consumption of the population in these countries, the demand for energy resources and the need for energy supply have risen. Therefore, the industrialised and developing countries are more and more interested in the regions with rich energy resources. The Caspian Sea Region, Russia, Iran, the Arab Peninsula and Africa are the main regions with rich energy reserves.[103] In this regard, Bull argues that *"the energy supply of countries in an era of growing material interdependence perceives the need for predictability and mutual responsibility on commerce, communication and energy issues. He states that cooperation among countries is a choice in the anarchic nation state system which, with growing material interdependence, may become a pervasive characteristic of world society."*[104]

Gas is the most productive among the energy resources. Its production is easier than oil and it does not need to be refined. Gas is a more reliable energy resource than coal and oil; it is non-toxic, and does not include any detrimental substances. Apart from that, unlike oil and coal, gas does not require storage. It is also used in electric production and in various industries other than its domestic use in residences and use for central heating purposes. Recently, the share of gas in world energy consumption has been increasing. Twenty-two percent of the world's energy consumption depends on gas. On the other hand, the use of gas in electric production has remarkably increased. Offices and households especially use excessive amounts of gas for heating purposes. Particularly, heating constitutes a 75% share in gas use.[105] Additionally, world gas use is rapidly increasing and the share of gas consumption among world energy resources is going up. Gas consumption is expected to reach 4.72 trillion m^3 by 2020. While gas production in 2010 reached the level of 2.92 trillion m^3 with a 3.1% increase, in 2011 gas consumption gradually increased by 2.2%. In developing countries such as Asian countries and South and Central American countries, a huge increase

---

103 **Ibid.**
104 **Brown, Seyom (1996);** International Relations in a Changing Global System Toward a Theory of the World Polity, Westview Press p. 30.
105 **Eray, Aynur (2002);** Enerjide Tutumluluk ve Verimlilik, Temiz Enerji Vakfı Yayınları, (Energy Efficiency and thriftiness), Istanbul, pp. 4–5.

in gas demand is expected. Furthermore, the use of gas in electric production is increasing and by 2020 it is expected to reach 33% of total gas consumption.[106]

In the last 20 years, the world's gas reserves increased by 100%. There are ten countries with large gas reserves; among these, the Russian Federation owns the largest one with 44.8 trillion m^3 of gas, followed by Turkmenistan, Nigeria, Ukraine and some Arab countries. Especially in the African continent, in the region containing Algeria, Egypt and Asia Pacific, there has been a significant increase in gas reserves. According to the BP data, by the end of 2011, world gas reserves reached 208.4 trillion m^3. On the other hand, when we look at the regional distribution of gas resources, we can see that gas reserves expand to a wider area than oil. The Middle East, for example, owns 56.6% of oil reserves, yet owns 40.5% of gas reserves. Caspian Basin oil and gas reserves are the primary source of income for the Central Asian Turkic Republics, especially Azerbaijan, Kazakhstan, Turkmenistan and Uzbekistan. Some regions with oil reserves own the larger part of gas resources. The Caspian Region meets 6% of the world's gas need with its gas reserves. In addition, the region has met 4% of the world oil production until 2011.

Figure 1: Gas Production and Consumption by Region

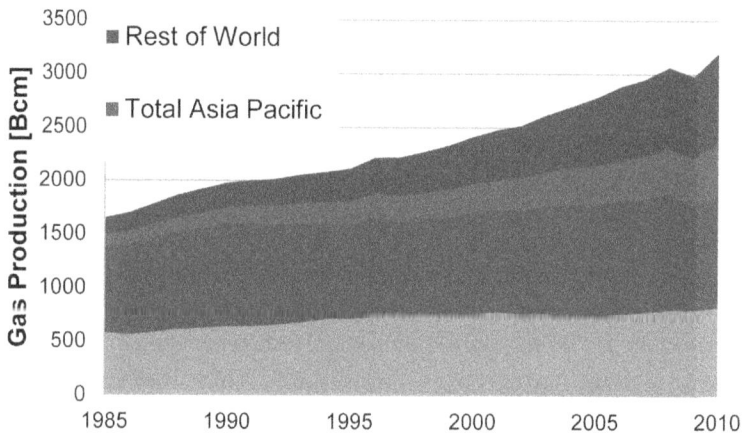

---

106 **Alemdaroglu, Nusret (2007);** Enerji Sektörünün Gelecegi Alternatif Enerji Kaynakları ve Türkiye'nin Önündeki Fırsatlar. (Future of Energy Sector and Turkey's Opportunity) İstanbul Ticaret Odası Yayınları, Istanbul p. 14.

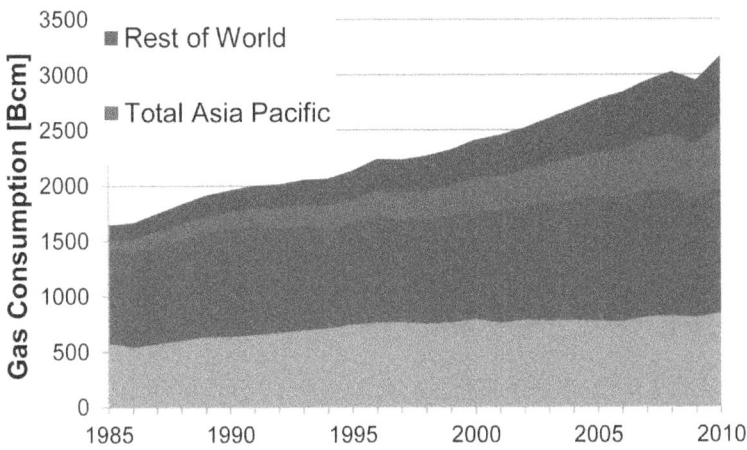

Source: BP Statistical Review of World Energy, June 2012.

As long as gas resources are used in accordance with today's level of production, it is estimated that we have enough reserves for the next 60–65 years. According to the BP data, world gas reserves reached 208.4m^3 and reserves are increasing in all regions. In terms of the total world gas reserves, the Russian Federation ranks first among countries with a share of 21.4% while the Middle Eastern countries rank first among regions. The total gas reserve of the Russian Federation is 44.6 trillion m^3, which is followed by Iran with a share of 15.9% and 33.1 trillion m^3. As the table indicates, gas consumption remarkably increases in the fourth quarter when the need for heating is higher. Especially in 2005, 2006, 2007, 2008 and 2009, when we look at the first quarter period, we see the remarkable change in Asia Pacific, which reaches 479 billion m^3. An overview of the world gas production since 1984 shows that it has reached an all-time high with a 7.3% increase. As is shown in the graphics, while the Russian Federation is the country with the highest increase in gas production, the Middle East stands out in gas consumption with a 7.4% increase. America is among the largest consumers of gas in the world.[107]

---

107 BP Statistical Review of World Energy, June 2012.

4.2 The Fossil Energy Resources used in the World and their Consumption

With the oil crisis, the delusive thought that oil is unlimited and cheap has disappeared and oil-based energy consumption is considered to be risky. The oil crisis of 1973 changed the energy policies and revived the alternative energy resources such as solar, wind and geothermal energy. After the crisis, states woke up to the fact that energy can be depleted and therefore should be used efficiently.[108] In the 1990s, the fluctuation of oil prices became a serious problem for both producers and consumers. The efforts to decide on oil prices were the focus of many discussions in the international arena.[109]

Figure 2: World Energy Consumption

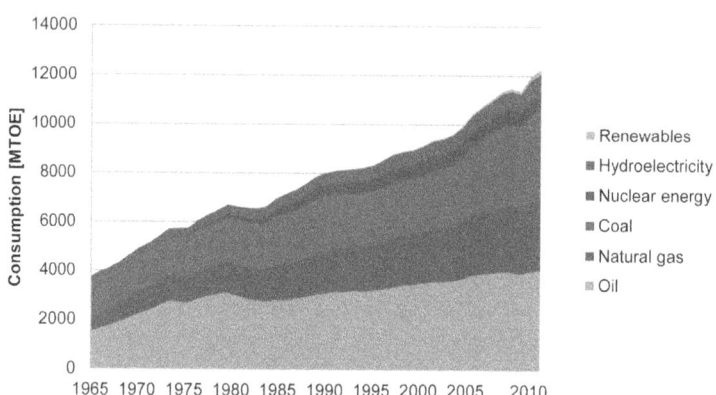

Source: BP Statistical Review of World Energy, 2012.

As it is seen in the above figure from BP, energy consumption in the world is constantly increasing with industrialisation. Most of the energy that we use can be supplied easily at affordable prices because they are acquired from fossil fuels with limited reserves such oil, coal and gas. Regarding the oil depletion issue,

---

108 **Yığıt, Ali (2001);** Elektrik Enerjisi Politikaları, Doğalgaz & Enerji Yönetimi Kongre ve Sergisi, Gaziantep. (Electricity Energy Politics, Natural Gas and Energy Management Congress), p. 234.
109 **Vander, Linde. C (2002);** The State and International Oil Market, Competition and the Changing Ownership of Crude Oil Asset. Boston/Dordrecht/London, Kluwer Academic Publishers, p. 22.

researchers estimate that we will be running out of oil reserves in 35–40 years, gas reserves in 65 years and coal reserves in 220 years. The following part of the report indicates that in the 2012 World Primary Energy Supply, oil has a share of 33.1% (600 000 b/d), coal of 30.3%, and gas of 20.9%, making it clear that the resources with limited reserves are coal, oil, and gas. On the other hand, the shares of nuclear and hydroelectric energy are 4.3% and 1.6% respectively. In the World Primary Energy Supply, the share of oil has decreased by 10.5% while the share of gas has increased by 5%. Moreover; there is a continuous increase in renewable energy resources and gas consumption in the world.[110]

Table of World Fossil Fuel Reserves

Region	Oil (million tonnes)	Gas (trillion m^3)	Coal (million tonnes)
North America	33.5	10.8	**245.0**
Central and South America	50.5	7.6	125.0
Europe and Eurasia	19	78.7	304.6
Middle East	108.2	80.0	328.9
Africa	17.6	14.5	
Asia and Oceania	5.5	16, 8	265.9
Total World	234.3	208.4	860.9
Former USSR Countries	17.2	74.7	228.03
OECD	35.7	18.7	378.5
OPEC	168.4		

Source: BP Statistical Review of World Energy, 2012.

According to BP data, the crude oil reserve in the world is 2343.7 billion tonnes. The largest oil reserve is in Middle Eastern countries with a share of 48.1%, which is followed by South and Central America with a share of 19.7%. The member countries of the Organisation of the Petroleum Exporting Countries (OPEC) own 72.4% of world oil reserves. The gas reserves of the world are found to be 208.4 trillion m^3. While the former USSR countries are ranked first with 35.8%, Iran comes right after the USSR countries with a share of 15.9%.

---

110   BP: Statistical Review of World Energy, June 2012.

We see that the largest oil reserves are in the Middle East while the largest gas reserves are in the former USSR countries. On the other hand, there is a more balanced distribution of coal and mineral coal resources in the world. As the statistics indicate, the Middle Eastern countries are becoming strategically more important due to their rich oil and gas resources. Former USSR countries are in the forefront in terms of gas reserves. Based on BP data, the Organisation for Economic Cooperation and Development (OECD) countries are dependent on the Middle Eastern and former USSR countries for both oil and gas supply. Most of the OECD countries, including the US and the EU, enable their gas and oil production with the gas and oil imported from OPEC member countries within the reserves of which they are in possession. Considering the prices and the current technology used, there are no problems with the supply in the middle term.[111]

This data demonstrates that, in the future, the US and the EU will be more dependent on the countries that have rich oil and gas reserves. As a result of the increase in gas use, the importance of the Caspian Sea region as well as Iran and Russia is increasing. Furthermore, dependency on the countries that are not members of the OPEC is gradually decreasing. The reason for this is that the Persian Gulf owns 60% of the total reserves.[112]

Figure 3: Fossil Fuel Reserves Production (R/P) Ratios in 2011

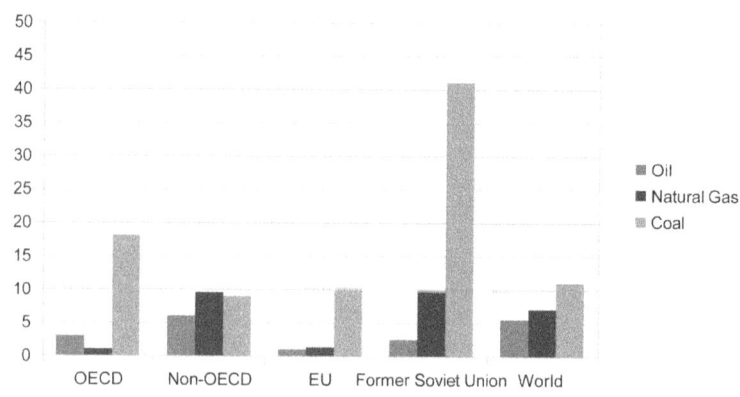

Source: BP Statistical Review of World Energy, 2012.

---

111 BP: Statistical Review of World Energy, June 2012.
112 **Aad, Correlje and Coby Linde (2006);** Energy Supply Security and Geopolitics: A European Perspective, Energy Policy. Vol. 34, Issue 5, March, p. 539.

In the 2012 BP statistics, we see that oil reserves have a wider distribution. While there are oil reserves in various parts of the world, gas is only found in certain countries and regions. Two thirds of the world's gas reserves are in Russia, the Caspian Sea and the Middle East. Therefore, the biggest gas consumers such as the EU countries, the US and Japan will be more dependent on Russia, the Caspian Sea and the Middle Eastern countries that are rich in oil and gas.[113]

On the other hand, the energy consumption in China and India is rapidly increasing. It is predicted that in the future, these countries will be dependent on the oil and gas reserves in the Caspian Sea region and the Middle East, just like the EU and the US, and most of their oil and gas needs will be supplied by these regions. In the world's energy supply, the fluctuation in demand is overcome with the reserve capacity allocated by Saudi Arabia in case of temporary loss of production. The OPEC, UEA and multinational oil companies play an important role in determining the oil policy in the world. In the Oxford Energy Forum *Why Oil Prices Have Moved Higher*, Horsnell states that "*if it were not for the stockpiles, the decrease in Iran's petrol supply in 1979 and the temporary loss of production in Iraq in 1980, 1990 and 2003 could not have compensated at all. Thanks to the stockpiles, OPEC countries did not suffer from any kind of oil shortage.*" However, Horsnell also thinks that the "*stockpiles in Saudi Arabia cannot meet the recently increasing energy demand due to the constant increase in the chronic energy demand since the 1990s.*"[114] Additionally, we see several political problems and instabilities in the oil-producing countries of the world. Due to the political instability of countries such as Russia, Iraq, Venezuela and Nigeria, these countries cannot give confidence to consumer countries in terms of providing sustainable energy supply in spite of the high amount of reserves they own.[115]

The countries of the EU are particularly dependent on the accessible countries such as Russia, the Persian Gulf, and Africa. Although Russia, Africa and the Caspian Sea region are positioned as oil and gas producer countries and are diverse alternatives of supplier countries for the consumers, the Persian Gulf, with its big potential and profound oil and gas reserves, is incompatible and the dependency on this region will increase in the future. Although these regions with rich reserves have no alternatives, they are also quite risky from a consumer country's point of view. The probable political and economic instabilities in these regions

---

113 BP: Statistical Review of World Energy, June 2012.
114 **Horsnell, Paul (2004);** Why Oil Prices Have Moved Higher, Oxford Energy Forum, P, 10–12 August, p. 10.
115 **Adellman, M. A. (1993);** The Economics of Petroleum Supply: Paper by M. A. Adelman, 1962–1993, Cambridge: The Mit Press. p. 34.

are potentially risky for the international partnership agreements. The income acquired from the oil and gas flow, employment, and consumption inducement function as public opinion pressure to create economic and political instability. When one looks at the regimes of the producer countries, we see that they have centralist regimes. Leite C. Weidman emphasises in his articles that *"there is no effective political authority in these countries"*, and *"the structures of civil society is quite weak"*. He thinks that *"the weak political authority and ineffective structures of civil society have a negative impact on the investments in creating oil production, stockpile investment, new production areas and transportation facilities"*.[116]

Since only a select part of the elite possess the income acquired from energy resources, the countries are more vulnerable to social decay and instability. The producing countries are not only reluctant about investment planned for improving production facilities and oil-gas transportation, but also about direct capital investment planned for creating new reserve capacity. Apart from that, projects such as TANAP-TAP, which will supply the EU with gas, are multinational projects and offer a convenient ground for direct capital investment. Yet, there is such limited investment in gas projects due to seasonal investment, distance and political risks.[117]

4.3 The Fluctuations in Energy Policy After 9/11

The Persian Gulf states own approximately 75% of the world's known reserves. In addition, the operational cost of oil is lower when compared to the other regions, which makes this region more advantageous. In return for the stabilisation of oil supply coming from the Persian Gulf, the security of the Western regions is ensured. In their book *"The Persian Gulf, Global Oil Resources, and International Security: Contemporary Economic Policy"*, Chapmann and Khanna point out that *"Iraqi occupation in Kuwait ended with military support coming from the US and Europe, and that these countries tried to balance the prices despite the possibility of increasing their share and income in the energy market."* In the 1980s and 90s, the stability of energy prices was regarded as a part of long-term strategies. The Persian Gulf states took lessons from the energy crisis in the 1970s. They learned that the stability of energy prices plays an important role in the sustenance of both demand and supply. Chap-

---

116 **Leite, Carlos/Weidmann, Jens (1999);** Does Mother Nature Corrupt? Natural Resources, Corruption and Economic Growth, IMF Working Paper, WP/99/85 Washington DC. p. 15.
117 **Morse, Edward Lewis (2002);** The Battle For Energy Dominance in Foreign Affairs (March/April), p. 81.

mann and Khanna state that *"because of the structural changes in the energy market, importer countries are concerned about the energy supply security".*[118] After the 9/11 attacks, the war in Iraq occurred in 2003, and the continuing instability in the region caused the US to be impotent in enabling security in the region, thus threatening the global energy supply's security in the region.[119] Furthermore, the Arab Spring, in addition to the earthquake and tsunami in Japan shocked energy markets, and lastly, the Libya revolution affected the loss of oil in 2011. The culmination of these events caused oil prices to hit an all-time record high.

Figure 4: Rotterdam Product Prices and US Gulf Coast Product Prices

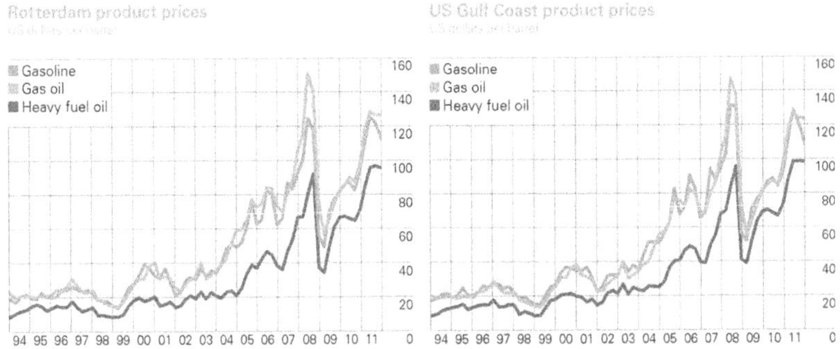

Source: BP Statistical Review of World Energy, 2012.

In the BP table, we can see the amount of oil and gas production. As indicated in the above table, these resources are continually increasing, and have been rapidly rising since 2001. After 9/11, energy supply security became the initial material of national and international agendas. By 2003, world oil prices had continually increased and in 2007, prices were doubled. In mid-July of 2008, while the price per barrel of oil went up to $147, in 2010 it decreased to the level of $90. There is an inconsistency between the world energy resources and the consumption speed. Since production cannot increase as rapidly as consumption, prices will go up accordingly and supply deficit will therefore gradually increase.[120]

---

118 **Helm, Dieter (2005a)**; The Assessment: The New Energy Paradigm. Oxford Review of Economic Policy, vol.21 No.1, Oxford University Press, p. 6.
119 **Chapmann Duane/ Khanna Neha (2006)**; The Persian Gulf, Global Oil Resources, and International Security. Contemporary Economic Policy, Vol. 24, Issue 4 October, pp. 507–519.
120 International Energy Agency:2012, World Energy Outlook, p. 1.

4.4 Rise of Natural Gas Use in the Electric Power Sector around the World

According to the 2010 Report of Electric Market data prepared by the Energy Market Regulator Report, world electricity generation is 20.181 twh, 41% of which is acquired from coal, 21.3% from gas, and 15.9% from hydroelectric resources. On the other hand, in the 2008 International Energy Agency (IEA)statistics, it is stated that in 2008, gas was mostly used in the generation of electricity. The distribution of global electricity generation according to the resources is as follows:

Figure 5: World Electricity Production in 2008

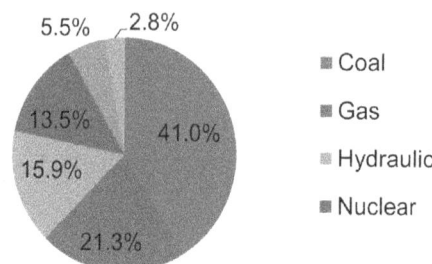

Source: International Energy Agency, 2010.

Apart from that, the share of gas in electricity generation in Turkey is more than twice as big as its share in world electricity generation. Moreover, the share of coal is below the world average. Its share in hydroelectric resource production is more or less the same. On the other hand, 41.7% of the electricity produced by 2008 is consumed in industry, while 1.6% is consumed in transportation and 56.7% is consumed in agriculture, commercial services, public services, residences and other areas. The biggest electricity producer is the US with 4,143 twh, while the biggest electricity exporter is France with 48 twh, and the biggest electricity importer is Brazil with 42 twh.

On the other hand, in 2008 the three biggest electricity producers were the US with 838 twh, France with 439 twh and Japan with 258 twh. France acquires 7.1% of its electricity generation from nuclear power plants. In terms of producing electricity from nuclear energy, Ukraine is ranked second with 46.7% and Sweden is ranked third.[121]

---

121 EPDK: 2010, Electricity Market Report of Turkey, pp. 1–3.

4.5 Scenario of World Natural Gas

World natural gas consumption grew in 2010 by 2.2%. However, global natural gas production grew by 3.1%. In this scenario, according to the report titled "An Overlook of World Energy" published in 2010 by the International Energy Agency, between 2008 and 2035 world primary energy demand will increase by 36%, but this increase will be slower than the last 27 years. In the last years, energy demand increased by 2%, but from now on it is estimated to decrease to the level of 1.2%.

Figure 6: Primary Energy Demand in the New Policies Scenario

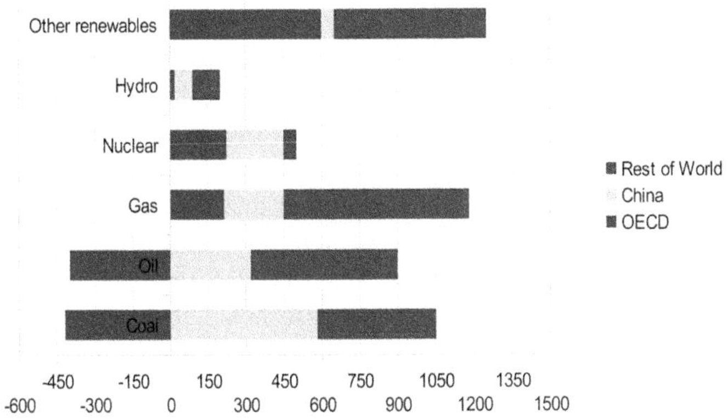

Source: International Energy Agency, World Energy Outlook, 2010.

It is foreseen that between the years 2008-2035, fossil fuels such as oil, gas and coal will preserve their dominancy and these resources will meet 75.7% of the energy demand in the following years. The rest of the energy demand will be acquired from biomass and waste (8.5%), renewable (6.6%), nuclear (6.4%) and hydraulic energies (2.8%). According to the scenario predictions, in 2020, it is foreseen that oil will have a 29.8% share in the primary energy supply, while between 2030-2035, oil will be replaced by coal, which will have a share of 29.11%. As for gas, it will maintain its importance in electricity generation and will have a share of 21.4%. In the case that existing policies are maintained, it is estimated that the fastest increase will occur in hydraulic and renewable energy resources (3.6%).

Figure 7: Coal-fired Electricity Generation by Region in the New Policies Scenario

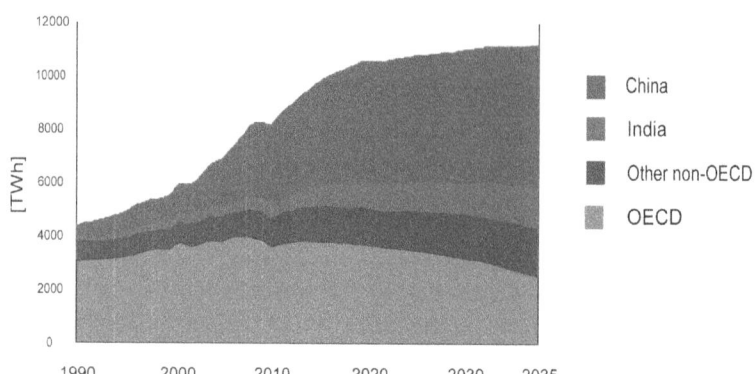

Source: International Energy Agency, World Energy Outlook, 2010.

Hence, it is predicted that the share of coal in world energy consumption will reach up to 29% in 2035. The reason why coal consumption and energy needs will increase is explained with the assumption that the consumption rate of developing countries such as China, India and Indonesia, which are not members of OECD, will significantly increase and in 2035, the share of coal in the electricity generation is estimated to reach up to 43%. It is expected that energy consumption will increase as a result of the growing population and continuing economic growth.

In 1973, the energy consumption of China in the world was 7.9%, while in 2008 it reached 16%. China used to be self-sufficient with its rich coal resources and limited oil reserves; however, as the recent growth was approximately 9% and recent energy demand increased by 70%, the country has become dependent on imported energy. The growth share of China in world oil demand is 8%. However, it has reached up to 30% from 2000 until the present. By 2035, China will be consuming 55% of the total energy supply. By 2035, it is estimated that China will be the biggest energy consumer country followed by India, the US and the EU; the remaining three biggest energy consumer countries.[122]

On the other hand, national energy companies play an important role in conducting strategies for national energy security. The fact that the national energy companies have taken their place in the energy market as the new players has

---

122 International Energy Agency: World Energy Outlook, 2010.

had a remarkable impact on the world energy supply security. These companies are backed by the national states and expand their investments in energy supplier states, which helps them to found new geopolitical relations and partnerships. Therefore, the national energy companies of China and India shape the oil and gas market that is dominated by western energy companies.[123] For this reason, other energy importer countries such as the US and EU countries feel that their interests are threatened. The other point is that since most of the countries with energy reserves are governed by anti-democratic regimes, whether these countries can meet the production increase and which prices they will demand is considered to be a risk for the security of the world energy supply.[124]

Another serious threat to the security of the global energy supply is the presumption that the dependency of the energy importer countries on the region will increase in the future. Although an oil boycott such as that of the 1970s is not expected, there is always the possibility that the OPEC will keep the prices high by using its market domination.[125]

4.6 Competition between Actors on Energy Issues

Energy is the source of economic development and social improvement. Thus, power is the determining factor for humans' welfare and state economy within the balanced and advanced development. The level of welfare in a country both depends on the per capita income and also on the per capita power consumption. A great competition in the world's energy field has been initiated through developing globalisation. Many Asian countries such as China, India and Japan are accepted as the regional blocks. The dependence on gas and petroleum of these three countries is rising day by day, and this situation makes cooperation essential for these countries. Likewise, the USA is in an endeavour to protect its own position against these block countries. Turkey has been an inalienable actor in the international and regional energy cooperation field thanks to its closeness to Eurasia, the Middle East and the Mediterranean in the geopolitical structure, its growing energy market, its reliability and stabilisation in the

---

123 **National Bureau of Asian Research (NBR) (2007)**; The Rise of Asia's National Oil Companies: Competitive and Geopolitical Implications. NBR Conference Report. p. 2.
124 **Kröger, Wolfgang (2006)**; Issues of Secure Energy Supply. Symposium, October 12, Zürich, p. 6.
125 **Birol, Fatih (2007)**; World Energy Prospects and Challenges. Asia-Pacific Review, Vol. 14, No.1, p. 3.

region, and its status as a NATO member state.[126] There were some threatening developments for Turkey during the Iran–Iraq war in 1980 by the Union of Soviet Socialist Republics' rising power and the foundation of pro-Iranian regimes. Thus, Turkey and the USA had a common interest during that time. The disruption of the petroleum trade that depends on the developments in that region has been a particular concern for Turkey as well as the EU and the USA.[127]

On the other hand, Iran's threats to Mosul and Kirkuk in 1986 have caused anxiety with regard to both Turkey and the USA. Moreover, Iran's occupancy of the oil fields in the region has been alarming for both countries. The USA's concern for Central Asia in the 1990s gradually increased. The balance in that region transformed just after the foundation of the Russian Federation by the dissolution of the Soviet Union and fifteen new states gaining their independence. Thus, a new strategy has started regarding the transportation of the rich underground resources in Central Asia and Caucasus to the Western and Eastern markets.[128]

Russia has been trying to prolong its efficiency in this region and is pursuant to near-abroad policy with the founded states within this field; it has also endeavoured to have the oil and natural gas pipelines pass through its lands. In addition, it desires to use, as a political trump card, the transit of the natural gas and oil exports from Central Asia to Europe and the world markets. Additionally, the USA accepted this region as an important matter for its strategic as well as its vital interests. Thus, the USA has aspired to cause Caucasia and the Central Asian countries to fall under the domination of Russia again, and to hinder the efficiency of Iran and China in this region. Within this scope, several American oil companies such as Exxon, Amoci and Chevron have entered into many agreements influentially to fill the voids raised in the Middle East just after the Cold War, and the USA has pursued a goal to improve its political efficiency in this region via Turkey. Moreover, it has also wanted to stop Iran's religious and political efficiency. In contrast, the USA tried to prevent the oil

---

126 Arzu Yorkan, Küresel Enerji Denkleminde Türkiye: **Available at;** http://www.bilgesam.com/tr/index.php?option=com_content&view=article&id=371:kuresel-enerji-denklemindeturkiye&catid=131:enerji&Itemid=146 (accessed on 28/12/2012).

127 **Oran, Baskın (Ed.) (2006);** Türk Dış Politikası II, 1980–2001, (Turkish Foreign Policy II 1980–2001), Ilhan Uzgel, ABD ve Nato, yla Iliskiler (Relations with USA and NATO) İletişim, İstanbul, p. 44.

128 **Blank, Stephen (2008);** The Strategic Importance of Central Asia: An American View, Parameters 38, No.1, Spring, p. 73.

and natural gas in the region from going down to the Gulf so that they could reach Ceyhan through Turkey.[129]

As for the USA and the European countries, they have tried to create alternative routes for Russia in the matter of the pipelines on the other side. Thus, they have aimed to minimise the efficiency of Russia and Iran in the region. The importance of the region has been addressed strategically for the USA's power policy, particularly after the September 11 attacks. It has been stated that the USA prefers the oil and natural gas of Turkmenistan, Uzbekistan, Kazakhstan, Iraq and Egypt to first reach Turkey and then be transported to the Western markets from there. Within this scope, the USA has been supporting projects like BTC, BTE and TANAP, which enable the transport of oil and natural gas from the Caspian region to Europe through Turkey.[130]

In a report prepared for the US congress, it is mentioned that Russia is disturbed by Turkey's rising efficiency in the region, and has started to engage with Iran closely, creating a new competitive environment through the partnership of the USA and Turkey against Russia and Iran. It is emphasised that Turkey increased its authority in the region by bringing forward the Stabilisation Covenant Project with the support of the USA at the end of 1999. The agreements that Turkey signed with Russia and Iran have lately been followed with deep concern exhibited by the USA. Turkey has been cooperating with the USA in many projects, but the USA felt discomforted by the natural gas agreement between Turkey and Iran in 1996 and the Blue Stream Project by Russia. It is known that the USA is against any kind of activities that would increase Iran's efficiency and strengthen it economically, such as the agreements signed with Iran. However, Turkey compromised with an Italian company to put the Blue Stream Project into practice just a week after the Baku–Ceyhan agreement in 1999.

On the other hand, Turkey has been discomforted by the USA's support to pipelines beyond the Balkans within the frame of the USA's multiple pipeline policy. Especially, there was a project that enabled the petroleum to come from the Caspian region to Russia's Novorossiysk port in the Black Sea. The petroleum was then going to be transported by the tank ships to the Burgaz port in Bulgaria. As for the other project, it was to ratify the Baku–Ceyhan pipeline due to the trans-

---

129   **Oran, Baskın (Ed.) (2006);** Türk Dış Politikası II, 1980–2001 (Turkish Foreign Policy II 1980–2001), Ilhan Uzgel, ABD ve Nato, yla Iliskiler (Relations with USA and NATO), İletişim, Istanbul, pp. 278–279.

130   **Amy Myers Jaffe/Ronald Soligo (2004);** Re-evaluating US Strategic Priorities in the Caspian Region: Balancing Energy Resource Initiatives with Terrorism Containment, Cambridge Review of International Affairs, Cilt 17, Sayı 2, p. 257.

portation of the petroleum via a new pipeline, which was being constructed to Dedeagac in the Aegean. This project has been a source of worry for Turkey, since this situation has been devaluing Turkey's trump card concerning the Straits.[131] As for Russia, it immediately developed the Nord and South Streams in order to not lose control of the transportation of the natural gas and petroleum on the other side. Russia has put these projects into action to transport the Caspian natural gas and petroleum by passing Ukraine and Belarus. China is another country that intends to be active in this region. The energy consumption in China has been increasing in the last few years due to its developing economy. Thus, China has made some pipeline agreements with Turkmenistan, Uzbekistan and Kazakhstan and developed some projects to take advantage of the region's petroleum and natural gas.[132]

---

131 **Oran, Baskın (Ed.) (2006);** Türk Dış Politikası II, 1980–2001: (Turkish Foreign Policy II 1980–2001), Ilhan Uzgel, ABD ve Nato, yla Iliskiler (Relations with USA and NATO), İletişim, Istanbul, pp. 280–282.

132 **Özkan, Gökhan (2010);** The Energy Security Dimension of Turkey's Regional Policy In The Central Asia And The Caucasus, Akademik Bakış Cilt 4 Sayı 7 Kış, 2010, p. 21.

5. EU Debates on Reducing Gas Dependency and EU-Russia Relations

5.1 Development of European Energy Policy

After the Second World War, European politicians assembled to form a common policy in the energy sector that would re-construct Europe. Firstly, on 9 May 1950, the seeds of the EU energy policy were planted with the declaration of France's minister of foreign policy, Robert Schuman: the European Coal and Steel Community (ECSC). This treaty entirely involved the German and French industries and, in particular, the common use of coal and steel in the war industry. Apart from that, the benefits of European participation in this economic community are mentioned.

The fact that France and Germany acted together in coal and steel production in this way was appealing to Belgium, Italy, Luxemburg and Holland, which is why the number of unions reached six and the first energy integration process started. The European Integration process extended with the ECSC, which was signed in the Paris Treaty of 1951, the treaty whose founding father is Schuman and where six unions gathered.[133] Thereafter, new mines were opened and more coal production was encouraged under the treaty. Furthermore, the basics of coal and steel production, consumption and pricing were determined. Since the treaty became effective in 1952, coal trade between the members of the union was active without any limitation.

The six countries that came together on 25 March 1957 in Rome founded the European Atomic Energy Society (EAES). After the ECSC, the success of the system that enabled free circulation for coal and steel products led the founding countries to cooperate in the area of nuclear energy. Additionally, the concept of energy was the focus of the economic cooperation that was founded by the EU in the 1950s. Philip Andrews pointed out that since this system worked well, the ECSC and the European Atomic Energy Community announced that West Europe needed energy urgently, and the substructure of a cooperation that would reduce war risk in Europe was thereby shaped, moving the economic focus of the European Community to another sector.[134]

---

133 European Commission: Summaries of EU Legislation, Treaty establishing the European Coal and Steel Community, ECSC http://europa.eu/legislation_summaries/institutional_affairs/treaties/treaties_ecsc_en.htm (accessed on 28/12/2012).

134 **Andrews, Philip (2003);** Energy Security in East Asia: a European View. Presented at the Symposium on Pacific Energy Cooperation 12–13 February, Tokyo.

The community work that started the ECSC treaty expanded to the European Economic Community in 1957 with the creation of a common market that would include sectors other than coal and nuclear energy. The customs tax between the member states, importation quotas and the obstacles in front of trade were aimed to be removed so that trade could improve with the formation of a common market.[135]

In his book titled *Transformation Analyses of the European Union*, Ibrahim Canpolat emphasises that the purpose of the European Atomic Energy Society is to provide wide research facilities in the field of energy among the member states. He thinks that "*the decline in oil costs and France's sidings with national control at that time undermined EAES's status*"; moreover, "*although nuclear power is economically competitive, it is laden with incremental costs.*" He states that the deconstruction of the old power plants incurs significant costs, let alone the construction of the new power plants. Considering the nuclear disasters such as Chernobyl, he says these risks reduced the prestige of the EAES at that time.

Furthermore, Canpolat mentions that since the European Economic Community did not include special provisions about the energy sector, the operation of raw oil, gas and electricity markets were not entirely stated. Therefore, when the EEC was formed, no decision was taken to develop a common energy policy. In the community activities, there were different responsibilities depending on the energy type. Canpolat states that while the European Atomic Energy Society is interested in splitting the atom for a peaceful use, the ECSC is responsible for the coal and steel sector. He considers the European Economic Community as a community taking responsibility for a wide energy policy including electricity, gas and oil.[136]

In the 1960s, the need for an energy policy was felt due to the setbacks in the field of energy, inciting the commission to prepare an energy policy memorandum in 1962.

A protocol was also signed and with the Treaty of Rome, energy was assembled under the authority of the EU. Thus, the first seeds for a common policy were planted in 1964.

---

135 European Commission: Treaty establishing the European Economic Community, EEC Treaty http://europa.eu/legislation_summaries/institutional_affairs/treaties/treaties_eec_en.htm (accessed on 28/12/2012).

136 **Canbolat, İbrahim (1998);** Uluslar Üstü Sistem AB Bir Dönüşümün Analizi (Analyses of the European Union's Transmission), İkinci Basım, Istanbul: Alfa Basım Yayım, p. 112.

With the protocol that the ECSC accepted on the issue of energy, goals began to be determined. With the combination of the communities in 1968, the commission accelerated efforts for a common energy policy. According to the commission, the obstacles in the way of energy trade were still there, which is why a common energy market was needed. If this market was based on the needs of the consumers and the necessities of the competitive pressures, the security of the energy supply would be provided. To realise this a sector plan should have been made, including the measures that would be taken to influence the investment strategies of the members after collecting the data. Additionally, measures should have been taken on issues like tax adjustment, technical obstacles, and state monopolies.[137]

Yet, the application of these proposals was demanding. The member states resisted the objectives in question. Since the expectations of the members differ from one another, there have been difficulties in forming the energy policy. To give an example, as France had good relations with Arab countries, it did not want to restrict itself with a common energy policy. As for England, it aimed to influence the decisions of the other community countries in this field as the only oil exporter member of the community.[138]

The main factor that made the member countries get closer to each other in time was their concerns about the security of the energy supply, bringing about the introduction of compulsory oil possession by a council instruction in 1968. Initially, with the European Coal and Steel Community and, six years later in 1958, the signing of the treaties that founded the European Atomic Energy Society and the European Economic Community, the newly formed energy policy gradually developed in the 1950s, affected by the expanding and deepening dynamics of the community.

5.1.1 After the Oil Crisis

In 1973, members of the Organisation of the Petroleum Exporting Countries (OPEC) raised oil prices by 60% in order to increase the oil income, an increase that continued for years to come. The price per barrel was 8.3 dollars in 1974,

---

137 Ege, Yavuz (2004); Avrupa Birliği'nin Enerji Politikası ve Türkiye'nin Uyumu, (European Union Energy Policy and Integration of Turkey), 1.B, Ankara, UPAV, p. 12–13.
138 Canpolat, Ibrahim (1998); Uluslar Üstü Sistem AB Bir Dönüşümün Analizi (Analyses of the European Union's Transmission), İkinci Basım, İstanbul Alfa Basım Yayın, p. 194–196.

but in 1977 it reached 11.5 dollars, and in 1980 prices were 28 dollars per barrel. OPEC members also increased oil prices by 60%.

The increase in the raw material and oil prices in many countries and the results of the oil crisis also affected inflation. While the 1973-1974 oil shock made a significant impact on Europe in terms of economic conditions, the US was not affected in the same way by this crisis because it was also an oil-producing country. Europe dominated energy resources with low costs in the colonisation period but with the oil crisis, the need to develop a new strategy was felt after realising the dependencies in energy supply.[139]

The oil depression and environment issues that were popular in the 1980s were at the top of the agendas of the member countries, creating a need for a strategy that would protect the energy supply from external shock. In the 1980s, the EU adopted a policy that was meant to liberalise the energy market, and steps were taken among countries in order to open the domestic energy market and current markets to competition. In a meeting in October 1972, the state and government presidents of the EU mentioned the need for supplying continual and secure energy. The studies that were initiated by this meeting are regarded as the first and most important documents of the EU in terms of council decisions.[140] The nuclear power that many EU countries used with military purposes during the Second World War began to be used to develop energy, due to the increase in oil prices after the oil crisis. Energy density increased by 25% between 1973 and 1986, thereby decreasing energy dependency from 65% to 45%. While oil consumption in 1973 was 63%, it dipped to 47% in 1986. In electricity generation, the amount of oil and gas usage decreased to 15% from 40%. Between 1973 and 1986, there was a 70% increase in the energy production of the union, whose energy is acquired from oil and nuclear power plants. The electricity generated by nuclear energy meets more than one third of the total electricity generation, and the use of nuclear power has increased six times. This amount in the nuclear power equals 13% of total energy consumption. The new energy policy that the union established took effect and the energy dependency declined to a certain extent.[141]

---

139 **Canpolat, Ibrahim (1998);** Uluslar Üstü Sistem AB Bir Dönüşümün Analizi, İkinci Basım, İstanbul: Alfa Basım Yayım, p. 194-196.
140 **Cüneyt Yenal Kesbiç/Hamza Şimşek (2001);** Avrupa Birliği Ortak Enerji Politikası, (European Union Common Energy Policy), Muğla Üniversitesi SBE Dergisi, Sayı 5, p. 4.
141 **M. Hakan, Keskin (2007);** Genişleme Ve Derinleşme Süreçlerinde Avrupa Birliği Politikaları, (EU Policies on Process of Enlargement), Stratejik Araştırmalar Dergisi, Genelkurmay Askeri Tarih Ve Stratejik Etüt Başkanlığı Yayınları, Sayı. 9, Yıl. 5, p. 71.

5.1.2 European Union Energy Policies Between 1986 and 1995

In the council dated 19 September 1986, European countries formed the purposes of energy policy and formulated goals to be reached by 1995. The council decided that the efficiency of the energy demand would be increased by 20%, the share of oil consumption within the energy consumption would be reduced to 40%, the net oil importation would be reduced to less than one third of total energy consumption, the role of gas would be maintained in the energy balance, the consumption of solid fuels would be encouraged, the share of hydrocarbon would be reduced to less than 15% in electricity generation, the nuclear facilities, which have an important role in the energy supply of the union, would work in accordance with the security conditions, and finally, that new and renewable energy resources, especially hydroelectric power, would be developed. In line with these purposes, the council wanted to provide supply security and reduce the risks in the energy market, which were also present in the oil market, to a minimum. They also sought to reduce the risks of abrupt price fluctuations, diversify the energy resources, apply the pricing principles of the EU to all consumer sectors and all energy types and enable co-ordination in the relations of the EU energy sector with other countries.[142]

These objectives targeted in the 1986 energy policy, however, were not achieved due to both the Gulf War and the developments in the energy market, resulting in the need for a new energy policy. In 1993–1994, a new energy policy began to be practiced. Three main objectives were determined in this energy policy. Priorities such as sustainable development, supply security, and competition in the union were mentioned. Within this scope, important steps were taken for the energy policy in the form of publications; first a green paper and then a white paper titled *"An Energy Policy for the EU"*. The green paper was prepared by the EU commission in order to open up certain issues for discussion on the level of the EU, and to ripen the subject by making the parties produce ideas. By definition, a green paper is a goodwill gesture meant to encourage debate and is not binding. However, if it turns into a white paper, it can be binding and provide opportunities for law and legislation development. In this white paper, three objectives other than energy were mentioned. Issues such as increasing the standard of living, creating employment and improving solidarity between regions were the aims of paper. Apart from this, the paper mentioned the three founding treaties; the EU treaty contributed to monopolising the authority of the EU. In this way, the EU moved

---

142 Official Journal of European Communities, http://eurlex.europa.eu/LexUriServ/LexUriServ.do?uri=OJ:C:2009:221:FULL:EN:PDF (accessed on 16/01/2012).

ahead of the national policies, the obstacles to trade were eliminated and the EU became more active in the decision-making processes.[143]

The treaty signed under the European Energy Charter was revised and expanded with the participation of more than 50 countries, including the US and Japan. Thus, the EU supported the securing of the energy supply for the necessities of the industry as well as finance and production to modernise the energy resources of the oil- and gas-producing countries. In addition, member states fulfilled their responsibilities of the international co-operation and increased investments in the field of energy. The Energy Charter Treaty is the first multilateral document supporting international cooperation in the energy sector.[144]

In 1993, the programme *"European Community Politics and Action Programme for Environment and Sustainable Development"* was founded. In this programme, content is determined by thinking of different scenarios regarding the future of energy sector. According to these projections, the programme stated that, compared with 1993, demand would increase by 24% and carbon emissions will increase by 20% in 2010. Based on these scenarios, cautions and policies were developed.[145] With the approved programme, sustainable development, the integration of environmental precautions with other political fields, industry, energy, agriculture and tourism as well as the effects of these main sectors on the environment became the focal point. The programme encouraged the EU states to act together with the member states due to the fluctuations and problems in the energy prices and supply. Thanks to the policies adopted as a result of common action, the dependency on energy declined. Yet, in the 1990s, the dependency on export energy reached 50% due to the increase in domestic consumption.

5.1.3 Energy Framework Programme

In 1996, there was a draft proposal prepared by the commission for the common objectives of energy policy, and the Energy Consultative Committee was founded. Apart from that, there was a structure called the Directorate General of Energy Transport that pursued energy developments. According to the reports published in 1997, there was a capacity deficit in the oil refinery sector that needed a revi-

---

143 **M. Hakan, Keskin (2006)**; Stratejik Açıdan Avrupa Birliği Enerji Politikası Ve Uluslararası Güvenlik Sistemine Etkisi, p. 104.
144 **Baklacı, Pinar/Akıntürk, Esen (2006)**; Enerji Şartı Anlaşması, Dokuz Eylül Üniversitesi İşletme Fakültesi Dergisi, Cilt. 7, Sayı. 2, p. 98.
145 **Dura, Cihan/Hariye, Atik (2003)**; Avrupa Birliği Gümrük Birliği ve Türkiye (EU Customs Union and Turkey), Ankara: Nobel Yayınları, p. 275.

sion, and there was an increase in the investments made in gas infrastructure. Just after the annual reports, the commission gained official status in 1997 with the Energy Framework Programme to put the new energy tools into practice. In this project—called EFP—the policies and precautions about preventing climate change in the EU were evaluated under the title of "environment." The details of the precautions to be evaluated with the priorities as well as the important objectives on the issue of the environment are mentioned in the Environmental Action Programme. Furthermore, between 1998 and 2002 six different energy projects were created and an EFP programme regarding providing energy supply security, environmental protection and market liberalisation was developed in order to promote competition. ETAP was one of these projects, and was founded on 14 December 1998 with the council decision and includes a range of action and energy precautions concerning the energy sector. The content of the project foresees that supply security will be provided, the environment will be protected, the competition on the free market will be fostered and analysis and evaluation studies will be made in the field of energy.[146]

Another project is CARNOT. Developed in 1998, it encourages clean and efficient use of solid fuel. The aim of this project is to limit emissions, including carbon dioxide emissions and acquire cutting edge technology at the most affordable prices. Three million EUR were spent on this project between 1998 and 2000. By 2003, the programme was included in the Smart Energy for Europe project. For the extension of clean and efficient solid fuel use at the level of Europe, the project cooperated with several institutions such as the International Energy Agency, the World Coal Institute, National Agencies and the World Bank. Additionally, the programmes that foster information and cooperation between communities at the international level and the actors in these communities, as well as the programmes intended for industrial cooperation, have received financial support.[147]

The SAVE project is, on the other hand, another project that encourages energy saving, proposing that energy should be used efficiently in industry, trade and transportation. Within this framework, there have been informative policies that encourage energy saving at the level of communities. The first SAVE project was practiced between 1991 and 1995, while the second was practiced in 1996 and 2000. Furthermore, the Energy Framework Programme that covered the years 1998–2002 and comprised a five-year union strategy was unified. On 9 April 2002,

---

146 Energy Framework Programme (1998-2002) ETAP http://ec.europa.eu/energy/evaluations/doc/2002_energy_framework_program.pdf (accessed on 05/04/2013).
147 The Carnot Programme http://ec.europa.eu/energy/evaluations/doc/2002_energy_framework_program.pdf (accessed on 05/04/2013).

"Smart Energy for Europe 2003–2006" was accepted.[148] With regard to renewable energy, ALTENER is one of the most important projects that the EU has developed. With this project, the development of renewable energy resources and the reduction of carbon dioxide emissions were considered to be the primary goal.

With this energy policy, it is thought that new work areas will be created and the employment market will be boosted. With ALTENER, research and development activities, which would search for and encourage renewable energy resources in the community member countries, were supported.[149] For the period of 1998–2003, some priorities were determined to transport radioactive materials in a secure way. Priorities such as the smooth operation of the market economy, the revision of transportation arrangements, and technical support to candidate and newly independent countries were present. The European Commission for the SURE Project contributes in terms of support for feasibility studies, and financially supports the comparative analysis of the aspects that organise radioactive substance transportation in the EU countries and candidate countries. It also identifies the limits of radioactive substance transportation, the evaluation of radiologic risks and the avoidance of illegal transportation of radioactive substances in the EU countries.[150]

Another project, as part of the European Energy Framework, is SYNERGY. This project aims to develop a cooperative project that is interested in the external dimensions of the EU's energy politics. SYNERGY, designed as a cooperation project rather than technical support, aims to develop regional and cross-border cooperations. As part of SYNERGY, the EU financially supported several projects that supported the proliferation of the energy sector in countries that extended it to a diverse and large geography, such as the Mediterranean Sea, the Middle East, the Baltic Sea, and the Far East. The Haifa Conference for Cooperation in the Middle East also focused on cooperation and familiarising Mediterranean countries with energy policies, establishing the framework for legal and corporate investment in the energy industries in Europe, and maintaining the security of European investment in the energy sector in the former USSR countries. It also discussed the Black Sea Regional Energy Centre activities, the reconciliation of Europe and Russia in the energy field, Balkan energy power, and Baltic Sea energy

---

148 The Save Programme http://ec.europa.eu/energy/evaluations/doc/2002_energy_framework_program.pdf (accessed on 05/04/2013).
149 The Altener Programme http://ec.europa.eu/energy/evaluations/doc/2002_energy_framework_program.pdf (accessed on 05/04/2013).
150 Energy Framework Programme SURE http://ec.europa.eu/energy/evaluations/doc/2002_energy_framework_program.pdf (accessed on 05/04/2013).

task power. The EU-People's Republic of China energy cooperation conferences and Latin American and Asian countries carried out several projects on the issues of potential energy resources.[151]

5.1.4 White Paper: An Energy Policy for Europe

Prepared in 1995, *White Paper on Energy Policy* is a book considered to be a political document about the energy policy for the European Union. It analyses the future of energy until 2020. This white paper mentions the necessity of technology-based and innovative development in the energy sector that will contribute to sustainable energy. This study also talks about the significance of a competitive position of Europe in the market and in the growing global economic struggle for energy.[152]

The book emphasises that energy saving occurs with the efficient use of scientific and technological developments in energy reserves. The limits of the increasing demand for clean and efficient energy in Europe are also determined, and the domestic energy market is established as a focus for competition. In effect, a public policy that will aid in creating a convenient atmosphere for investments is formulated.[153] With the white paper, the liberalisation of the energy market in the EU is strengthened, as is the desire to unify the current markets, which are fragmented, between the countries. The paper was the first component of the EU energy strategy that was prepared to provide the European Union's domestic energy market.

5.2 The Green Paper (The New Energy Policy of the EU)

5.2.1 The Green Paper: A European Strategy for the Security of Energy Supply

Another important and basic book, and what could be considered the most comprehensive study that was prepared after the white paper for a common energy policy for the European Union, was the green paper titled *A European Strategy for the Security of Energy Supply*, written in 2000. In the green paper, the main issue related to energy supply security is mentioned succinctly: The EU's energy consumption and exportation of energy products are gradually

---

151 Energy Framework Programme SYNERGY http://ec.europa.eu/energy/evaluations/doc/2002_energy_framework_program.pdf (accessed on 05/04/2013).
152 **White Paper:** An Energy Policy for the European Union, http://europa.eu/documents/comm/white_papers/index_en.htm (accessed on 06/11/2011).
153 Energy in Europe: 1996, European Energy to 2020, European Commission Directorate General for Energy Brussels, Special Issue, p. 3.

increasing. Community production is insufficient in terms of meeting the energy needs of the Union. As a result of this, energy dependence in other countries is gradually increasing. If no measures are taken, 70% of the energy the EU needs will depend on imports in the forthcoming 20–30 years. This dependency can be seen in all sectors of the economy. When the geopolitical distribution of energy provided from abroad is considered, by 1999—a year before the publishing date of the green paper—it was observed that 45% of imported oil came from the Middle East and 40% of the imported gas came from Russia.[154]

Frank Schimmelfennig points out that *"after this strategy, the EU tended towards the eastern and southern. The commission decided to form an attitude for the union and to use economic and political power against the main supplier states."* He thinks that *"to apply pressure on these states is one of the main factors in the EU energy policy."*[155]

Considering the sustainable growth and environmental concerns in accordance with the Lisbon Strategy and the Kyoto Protocol, the first concrete step to provide the energy supply of the EU was taken with the green paper published in 2000. With this book, the energy supply security of the EU was, for the first time, discussed comprehensively. Thus, it was mentioned that bringing growth under control and managing import energy dependency were the primary issues of the future. Since there was no dissent between the EU countries on the issue of energy, it was stated that no sufficiency or movement area were provided in the process of negotiation and pressure. It was expressed that the EU energy supply security required a lot more than simple cooperation and the decisions made needed more power. Thus, physical, economic and social risks were also defined regarding energy supply security.[156]

5.2.2 Sustainable, Competitive and Secure Energy for European Strategy (2006)

This book was revised and renamed as *A common energy policy for Europe* at the end of 2005 with the gathering of the heads of the European state and governments

---

154 **Green Paper (2001);** Towards a European Strategy for the Security of Energy Supply, Brussels European Commission, p. 3.
155 **Lavenex, Sandra/Schimmelfennig, Frank (2007);** "Relations with the Wider Europe", Journal of Common Market Studies Annual Review of the European Union Volume 45, Issue 1 pp. 143-162.
156 Green Paper: European Commission, 2001.

in Hampton Court. The efforts of the European Commission were published for a second time on 8 March 2006 under a new title: *Sustainable, Competitive and Secure Energy for European Strategy*.

In the green paper, it was emphasised that the Union should diversify its energy supply since it is increasingly becoming dependent on external resources, and according to the scenarios, it is foreseen that this dependency will reach up to 70% in 2030. Furthermore, the book highlights that since 42% of the Union's energy needs are met by the Russian Federation, the Caspian Sea energy resources are considered to have great potential. On the other hand, the book says that the construction of transmission lines in the Caspian Region that will transport oil and gas to Europe are inevitable. It also positions countries on this route, such as Turkey and Ukraine, as a bridge in the energy transportation between countries. Therefore, the development of gas and oil transmission projects by the EU and the diversification of the energy supply resources rose to the top of the agenda.[157]

While the green paper was revised in 2006, it had been predicted that by this year, the EU would be dependent on three countries (Russia, Algeria and Norway) and in the forthcoming 25 years, the gas importation will reach up to 80%. The import dependency on energy, the increase in oil and gas prices, and a related increase in electricity prices are also underlined.[158] In summary, these issues are mentioned in the 2006 revised green paper:

– An entire improvement of the gas and electric market within the Union.
– Revision of the policies about the management of oil and gas stocks.
– The improvement of network security.
– The formation of a solidarity mechanism for the purpose of supporting member countries having problems with the infrastructure of the energy network that will be established throughout Europe.
– The initiation of union-wide discussions and search the for solutions about diverse energy resources and climate change.
– In accordance with the common energy policies, the formation of an Action Plan.
– The preparation of a roadmap for the renewable energy resources.
– A strategic plan for energy technologies.

---

157 **Green Paper (2006);** A European Strategy for Sustainable, Competitive and Secure Energy, Brussels, Commission of the European Communities, p. 25.
158 **Green Paper (2006);** A European Strategy for Sustainable, Competitive and Secure Energy, Brussels, Commission of the European Communities, p. 3.

- The preparation of a list of foreseen priorities in the Union in line with the foundation of a common energy network for the Union.
- The privileging of the EU-Russia Energy Dialogue and the completion of the Energy Charter Treaty.
- The initiation of efforts to make a Pan-European Energy Community Treaty that is based on the South-East European Energy Community.
- The development of a new union mechanism for the purpose of giving a fast and coordinated common answer in case of an emerging problem concerning the provision of energy resources outside of the EU.
- Aiming for a decrease in energy consumption in 2020 as opposed to today's consumption, so that energy savings worth €60 billion can be made, and yearly savings per household will be approximately €200–1000.
- Creating one million employment opportunities with energy savings.
- As it is foreseen in the Kyoto Protocol as well, aiming for a decrease in CO_2 emissions, with a 20% decrease in energy consumption and pollution reduction.[159]

The European Commission made important decisions about the EU members' common energy policy based on the advice in this book. According to the book, the goal of the liberalisation of gas and electricity was set, as well as goals such as accelerating relations with the main dealers like Russia and OPEC. Moreover, developing renewable energy resources and research on low-carbon technologies were also issues to address. The annual oil score is €250 billion in Europe. This amount is equal to the gross national product grip of the EU. Apart from that, the 2006 conflict between Russia and Ukraine and Russia's taking advantage of oil as a political tool unsettled the EU, influencing the EU to search for alternative energy.[160] The difference between the new green paper and the green paper published in 2001 is that the former puts an emphasis on a common energy policy and a common external energy policy for a more coordinated and efficient attitude towards energy supply security.

---

159 **Green Paper: 2006**, A European Strategy for Sustainable, Competitive and Secure Energy, Brussels, Commission of the European Communities, http://europa.eu/documents/comm/green_papers/pdf/com2006_105_en.pdf (accessed on 05/04/2013).
160 **Gawdat, Bahgat (2006)**; Europe's Energy Security: Challenges and Opportunities, International Affairs, London, 82:961–975 September.

5.3 Europe Energy Charter and Europe Energy Charter Treaty

The Europe Energy Charter laid its foundations in the 1990s. It started in June 1990, when the Prime Minister of Holland, Ruud Lübbers, came up with a proposal for the construction of the European Energy Community that was presented in the European Council meeting. In 1991, the declaration for the Energy Charter was signed in The Hague. The European Energy Charter that started with these negotiations began to be debated simultaneously with Uruguay Round and influenced by the rules that governed the World Trade Organisation. Thus, the European Energy Charter became the first important political document of the European Economic Community in the energy field, which organised the energy supply and demand conditions. In December 1994, three years after the Declaration of the European Energy Charter, the European Energy Charter was signed. At this time, the states that came together in the European Energy Charter conference signed the Energy Charter Protocol on Energy Efficiency and Related Environmental Aspects (PEERA) at the end of the conference, along with the Energy Charter Treaty. The ECT and PEERA went into effect in April 1998. The Energy Charter Treaty is based on the 1991 energy charter declaration. In this case, a simple goodwill gesture to foster energy cooperation turned into a binding and multilateral treaty.[161]

Providing cooperation on the issue of energy was a serious priority in comparison with several other sectors. In a world where dependency between countries has increased, the dependency of importer and exporter countries on each other increased as well. The deficiencies and problems in the atmosphere created by illegal tools and bilateral agreements pointed out the need for an agreement. In providing global energy security, the European Energy Charter came up with open, transparent, competitive markets and a legal structure based on sustainable development. Thus, with the Energy Charter Treaty, international rules were established between energy exporter and importer countries.[162]

This treaty determines the main principles of the energy sector and encourages investments and trade in this sector. It also enables secure energy transit, thereby increasing the efficiency of the energy and developing impartial mechanisms to solve the controversies in the sector. The purpose was to develop relations between the east and the west, namely European countries, former USSR countries, Central

---

161 The Energy Charter Treaty: A Readers Guide http://www.encharter.org/fileadmin/user_upload/Publications/ECT_Guide_ENG.pdf (accessed on 18/01/2012).
162 The Energy Charter Treaty and Related Documents (2004); A Legal Framework For International Energy Cooperation, Brussels, Energy Charter Secretariat, p. 13.

and Eastern Europe, Canada, the US and Japan, in the issues of the transportation of energy products, trade, investment and energy cooperation.

The Energy Charter Treaty seeks to minimise the risks in energy issues such as energy trade and energy investments in the international arena. This goal is shaped around the five main objectives given below:

- Encouraging and protecting foreign energy investments based on the most favoured national maxim.
- Free trade of energy materials, products and equipment in accordance with the WTO rules.
- Free transit pass of energy in the pipelines.
- Increasing energy efficiency and reducing the environmental risks of energy cycle.
- Developing diverse mechanisms to solve the controversies between the two states as well as between the host country and investor country.

Eduardo Benito emphasises that "*the process initiated by the EU with the attempt of European Energy Charter supported the development and reforms in the energy sector in the Russian Federation, especially Eastern Europe.*" He also points out that some conveniences are provided concerning the transit passes for the energy security, namely for the increasing transportation of energy resources from the producer to the consumer through international borders.[163]

Apart from that, according to the Energy Charter Treaty, preparations for the Transit Protocol that gathers clauses that include more details about transit passes started in 2000. The purpose of the protocol is to determine principles that would organise the transit pass of the energy resources including hydrocarbon and electricity.[164] The purpose of the provisions that organise the transit regime with this protocol is to overcome the legal and technical difficulties that former USSR countries face while transporting their energy resources with pipeline projects. The draft presented on 10 December 2003 at the Energy Charter Conference was accepted as the Transit Protocol text.[165]

The European Energy Charter Conference is an institution formed during the process of the EU energy charter. The Investment Group, Transit Group,

---

163 **Sodupe, Kepa/Benito, Eduardo (2001)**; Pan-European Energy Co-Operation Opportunities, Limitations, and Security of Supply to the EU, Journal of Common Market Studies, Vol. 39, Issue 1, March pp. 169–170.
164 The Energy Charter Treaty: A Reader's Guide p. 15.
165 Transit Protocol (accessed on 07/11/2011) http://www.encharter.org/index.php?id=37.

PEERA Group, Budget Committee and Legal Consultancy Committee study the issues of energy investment foreseen by ECT and PEERA provisions, trade, energy activity and environmental issues. Apart from these study groups and committees, the Legal Advisory Task Force (LATF) founded in 2001 is responsible for the operation of the Model Agreements Project for the transboundary oil and gas pipelines. In the near future, LATF is expected to develop projects about electric web transboundary pipelines. Pipeline Model Agreements are called Intergovernmental Model Agreements among states and Host Governmental Model Agreements between host states and the investor. The goal of the developed model agreements is to help facilitate negotiations on pipeline projects among countries and between the host and the investor, based on the ECT provisions.[166]

5.4. Security Risks and EU Policies on European Energy Supply

A leading regime theorist, Robert Keohane, focuses mainly on international economic cooperation. This idea is sustained on the basis of rational calculations of self-interest by national governments. He puts forward that the economic field takes precedence over the military. He also articulates that nation states are interested in subscribing to rules of international behaviour, but sometimes impose on domestic policies with strong international effects.[167] After World War II, a mutual dependence began growing among countries, and the interdependence approach is useful in pointing out any crises or political uncertainties that affect the countries. Firstly, the security of the EU's energy supply became a matter of debate with the petroleum crisis in the 1970s. Political uncertainties such as the Soviet Union breaking up, the Gulf Wars and September 11 affected the energy market and its prices. In 2003, reasons such as stopping imports from Venezuela disrupted the energy supply.[168] Terrorist attacks, wars, and disasters such as hurricanes, earthquakes, and tsunamis are also factors that affect energy supply and provision. The USA Energy Institute

---

166 First Edition of the Model Inter Governmental and Host Government Agreements for Cross Border Pipelines, http://www.encharter.org/index.php?id=182 (accessed on 07/11/2011).
167 **Brown, Seyom (1996);** International Relations in a Changing Global System Toward a Theory of the World Polity, Westview Press, p. 32.
168 **Daniel, Yergin (2005);** Energy Security and Markets, CERA in Energy and Security: Toward a New Foreign Policy Strategy, J. Kalicki and D. Goldwyn, Editors. 2005, Woodrow Wilson Center Press: Washington, DC.

splits the security of energy risks into three: first is the attack upon the energy systems; second, an attack by an energy system, for example the usage of chemical and biological materials for the energy plants; and lastly, the attack through the systems of energy, for example an attack that uses chemical and biological materials on transit pipes.[169]

According to the information given in the reports of the IEA (International Energy Agency), the world's energy supply security has faced different risks at different times. Events such as socialisation after the Second World War, the Suez Canal crisis, the embargo of Arabian countries, the Iranian revolution, the War of Iraq and Hurricane Katrina in the Gulf of Mexico decreased 20% of petroleum production and 16% of the EU's gas production, thus disrupting the production of some countries for extended periods. In addition to this, energy supply was cut on local and natural factors. Presently, reasons such as political uncertainty, civil war and terrorism in the countries that export petroleum are more likely triggers of this uncertainty. Energy supply risk is not only limited to the Middle East. It can also be seen in countries such as Indonesia, Africa and Latin America. These uncertainties increase the prices of petroleum and energy; for this reason, the IEA obliges the member countries to possess a 90-day reserve of crude petroleum in case of a disruption to the energy supply.[170]

Cankıran and Oktay assert that significant developments in the common energy policy of the Union are difficult to bring about in the short term. As an example of this, they indicate that Germany and France do not lean towards liberalisation in the energy market, and the member countries are not willing to delegate their authority to the EU because of their worries about national security. Some member countries' private agreements with Russia are examples of the nonexistence of a complete union on the policy of energy.[171] In the study that Daniel Steinvorth prepared for Friedrich-Ebert-Stiftung, he criticises this as follows: *"European Union needs a common energy policy. Energy import of this Union whose population is 500 million increases progressively and the EU will import 70% of its energy requirement from abroad 20–30 years later. Risks of energy supply should be ended, liability of Kyoto Protocol should be carried out and there should be a competition. Political differences should be ended between the member*

---

169 United States Energy Association, National Energy Security Post 9/11 (Washington, D.C.: United States Energy Association, 2002), p. 54.
170 **International Energy Agency (2003);** Energy Policies of IEA Countries Review.
171 **Oktay, E. /R.F. Cankıran, (2006);** Avrupa Birliği'nin Enerji Güvenliği Açısından Türkiye'nin Önemi, (Significance of Turkey in the Framework of EU Energy Security) Avrupa Araştırmaları Dergisi 14, pp. 153–173.

countries and especially Germany and France should act with their own national political decisions instead of the norms of the EU."[172]

By grouping countries together, common markets have been designed, as well as regional trading blocs and customs unions. It was planned that their domestic economies, hereby national autonomy, would be overcome by more economically powerful states or transnational corporations. The aim of the European Community is to integrate the market by shared perception among European Community members in order to compete against the United States and Japan; a strategy seen as a necessary correlation of international anarchy.[173] However, there are different opinions on energy policy strategies in the EU parliament. Alexander Graf Lambsdorf and Jorgo Chatzimarkakis, for example, prepared a document for the EU parliament energy group, stating that the EU needs a strong energy policy in this context, and that climate change, increasing energy dependence and prices threaten the security of the EU. In their analysis, they underlined that the energy policy should be integrated between the countries urgently, and in addition to this, there should be a single voice among the countries representing the external affairs policy of the EU's energy. The advantage of Brussels being the only addressee regarding the relations of the member countries with Russia and the Middle East was emphasised. As mentioned before, the energy dependence of the EU is 50% today, but is expected to reach 65% in 2030. Therefore, they recommended that especially the energy supply be diversified, and the TANAP and TAP projects be supported.[174]

According to the German Ministry of Economics and Technology, the EU is in the adaptation period with the new member countries, and is also making an effort to promote relations with contiguous countries. The Ministry expresses that a domestic market should be created, national authorities should be delegated to the EU, and therefore the EU should act indivisibly on the subject of external energy policy. The Ministry also stated that the EU's energy dependence on Russia is high, but this situation does not carry risk in terms of the security of energy.[175] Christoph Tönjes and Wilbur Perlot enumerate the same

---

172 Friedrich Ebert Stiftung-Frankreich, Info Nr. 3, 2005 by Daniel Steinvorth (Deutsche–französische Energiepolitik im europäischen Kontext), p. 3.
173 **Brown, Seyom (1996);** International Relations in a Changing Global System Toward a Theory of the World Polity, Westview Press p. 31.
174 **Graf Lambddorff, Alexander/Chatzimarkakis, Jorgo (2008);** Für die Gruppe der Europäischen Parlament Europäische Energiepolitik, Positionspaper.
175 Bundesministerium für Wirtschaft und Technologie: Energie in Deutschland, 2010 Trends und Hintergründe zur Energieversorgung, p. 27.

in their book: *"In case of crisis, EU countries should have an auxiliary reserve and sharing the auxiliary stock of the country having the reserve with the other member countries should be based on an agreement in the EU commission and this situation should be presented as reports."* With this recommendation, the effects of the extreme increase in petroleum prices will be reduced. Furthermore, they postulate that increasing the auxiliary petroleum reserve of the EU countries from 90 to 120 days is significant.[176] Apart from this, EU countries must have a natural gas stock that will be enough for a minimum of 60 days. Austria is the leading country in this subject with its 115-day reserve, and Denmark follows Austria with a 65-day reserve. Greece, England and Holland are among the countries whose reserves are minimal, with their natural gas reserves sufficing for 10–20 days. Surprisingly, some countries do not even have a natural gas reserve.[177] According to the EU Commission directives, all member countries must have an auxiliary reserve that will be enough for a minimum of 90 days, and the countries that have reserves for less than 90 days have to increase this period, as having an auxiliary natural gas reserve provides, in effect, energy security. On the other hand, increasing the energy efficiency of buildings and the transportation sector, and increasing renewable energy resources will reduce energy dependence on the energy resources outside the EU.

5.5 The EU's External Dependence and Energy Consumption

The EU is one of the regions where most of the energy in the world is consumed, yet it does not have enough potentialities in terms of energy generation or resources. In this context, the number of member countries became 27 after the fifth expansion, resulting in a considerable increase in terms of imported energy dependence. This situation makes new initiatives and approaches obligatory for the EU, because of the security of its energy supply. Some subjects, such as applying the multiple pipelines policy and thus diversifying the resources for energy imports, are among the EU's policy options on the priority list.

According to the report *The Energy Trend in Germany* from the German Federal Ministry of Economics and Technology, the country that consumes

---

176 **Robbert Willenborg/Christoph Tönjes/Wilbur, Perlot (2004);** "Europe's Oil Defences: An analysis of Europe's oil supply vulnerability and its emergency oil stockholding systems", The Clingendael Institute, The Hague, January 2004, p. 46.
177 Proposal for a Directive of European Parliament and of the Council – concerning measures to safeguard security of natural gas supply, 2002/0220 (COD), 2002.

the most energy is France, followed by Germany, England, Italy and Spain. The fuel on which EU countries are externally dependent is petroleum. Sweden is the country that is the least dependent on fossil fuels, as it benefits from renewable energy at a rate of 32% and nuclear energy at a rate of 33% for its energy supply. The most important reason for the countries' dependency in the field of petroleum is the high share used on transportation. On the other hand, geographic conditions are important for producing energy. As an example, when Swedish and Austrian hydroelectric potential is considered, it can be understood that geographical conditions are advantageous. On the other hand, France produces 76% of its electric energy from nuclear energy. It is predicted that dependence on fossil fuels in countries such as France and Germany will increase with the closure of the nuclear power plants. Ireland (87%) and Holland (86%) are the countries that have the highest dependencies on fossil resources.[178]

Figure 8: EU-27 Energy Import dependency

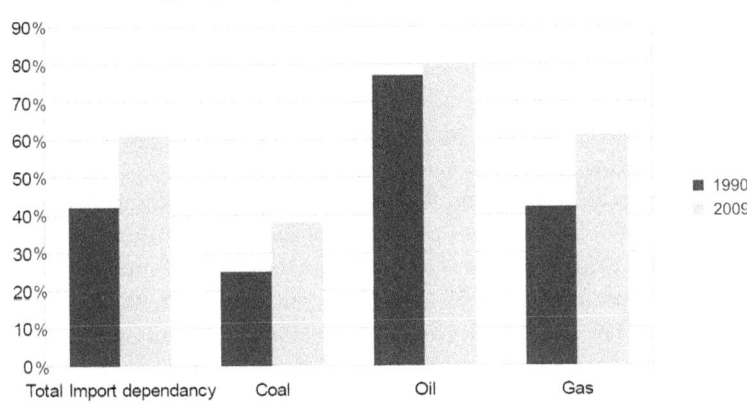

Source: Eurostat May 2011

The EU's energy dependence on fossil fuels has been progressively increasing, as can be seen from the table, reaching 83.5% for petrol and 64.2% for natural gas. The EU's energy dependence has been increasing with its expanding

---

178 **Bundesministerium für Wirtschaft und Technologie (2010);** Energie in Deutschland, 2010 Trends und Hintergründe zur Energieversorgung, p. 47.

structure.[179] It is externally dependent on energy at a rate of nearly 50%, but has to import two thirds of its petroleum; however, it uses its own resources for two thirds of its coal. The most important countries from where energy is imported are Russia and Kazakhstan, which possess gas and petroleum respectively. Petroleum is also imported from OPEC countries and coal is imported from North Africa, Colombia and Australia. According to the EU Commission's predictions, this energy dependence will only increase. With the increase in the external dependence on energy, the community ruled that petroleum and natural gas stocks are compulsory. Additionally, for the consistency of the energy supply, strengthening the supply transportation facilities also became important.

5.5.1. The EU and the Security of Natural Gas Supply

Natural gas imports in the EU are consistently increasing with developing technology. The market that imports the most natural gas is the EU, and this situation is expected to continue until 2030. Natural gas is thought of as the fuel that is least harmful for the environment among the fossil fuels, and for this reason it is usually preferred. Natural gas is also important in the struggle against climate change and in meeting the aims for a sustainable economic growth. According to data from 2009, natural gas is the second highest energy source consumed with a rate of 24.6%, following petroleum.[180]

According to the predictions, demand for natural gas will increase at a rate of 2.6% in the EU over the next twenty years, and an 80% natural gas demand is expected by 2030. The reserve capacity that the EU itself provides from the North Sea will decrease, contrary to the propensity to increase the consumption of natural gas. For this reason, the EU's dependence on imported natural gas is progressively increasing. According to data from 2009, the EU's dependence on imported natural gas includes a 34% import from Russia. Following Russia, Norway (31%) and Algeria (14%) are the other main exporters from which the EU imports natural gas.

---

179 Energy, transport and environment indicators, 2011 edition, Eurostat, European Commission.
180 https://ec.europa.eu/energy/sites/ener/files/documents/trends_to_2030_update_2009.pdf; p. 43.

Figure 9: Energy dependency – natural gas, 2009

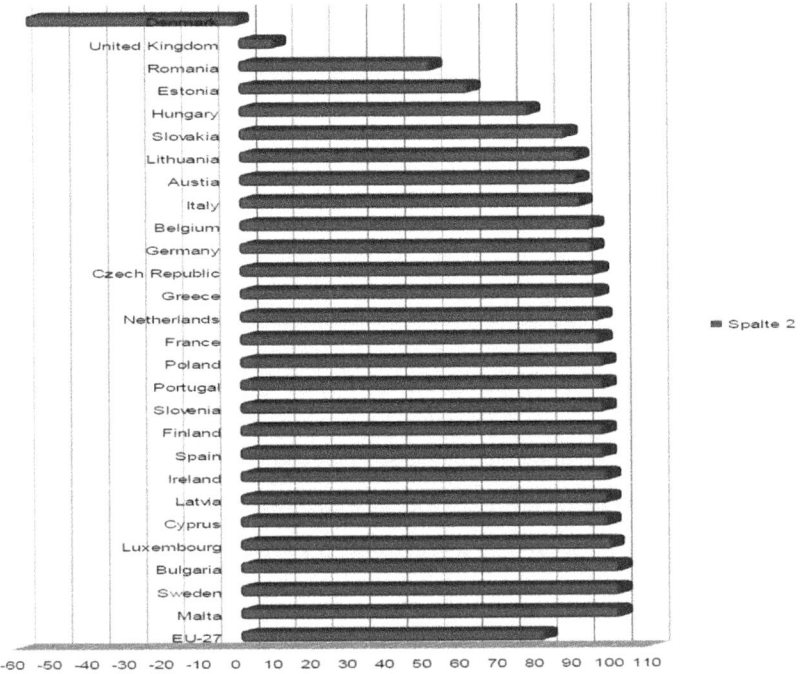

Years	1999	2000	2001	2002	2003	2004	2005	2006	2007	2008	2009
(%)	47.9	48.9	47.2	51.1	52.4	53.9	57.7	60.8	60.3	62.3	64.2

Source: Eurostat 2011

According to Eurostat's data, the dependence on imported natural gas increased by about 16% between 1999 and 2009. It is understood that this rate reached 64.2% in the EU in 2009. On the other hand, Holland and Denmark are the countries that are not dependent on natural gas, but are exporting natural gas through the EU. In the years between 1999 and 2009, Denmark doubled its export rate from 58.9% to 120.9%, and Holland increased its rate from 63.8% to 72.7%. As for England, while it was an exporter country until 2003, it began importing natural gas in 2004, and its import rate has reached

31.6%. Ireland increased its natural gas import to a rate of 92.5%, whereas it was 63.2% in 1999.[181]

5.5.2. The Effects of the Gas Crises on European Energy Security

Another assumption of international corporation is that countries need to organise a common defense against political enemies or economic rivals. Such competitive motives can be calculated through multilateral collaboration. Scientists were attempting to understand the origins and maintenance of such an international regime in the 1970s and 1980s. It is defined by prominent groups in international relations.[182] The EU consisted of regional blocks at this point, and decision makers and iron and steel cooperations with the European Union's six founder countries recently reached 27 member countries, a population of 500 million and the biggest economic power in the world with the expansion in 2007. According to the 2008 data, it constituted 30% of the world's gross domestic product. Moreover, when the purchasing power parity is considered basic, it has a Gross Domestic Product of 12.8 trillion and the income per capita is calculated at 25,900 EUR per year.[183]

This economic power is parallel with the European Union being the second biggest energy consumer in the world, in pursuit of the USA. Furthermore, EU energy consumption is increasing constantly. Seventeen percent of the energy consumption in the world takes place in the EU. Most of the EU's energy import consists of petroleum and natural gas. The European Union has experienced problems caused by imported energy dependence with the energy crisis of the 1970s. With this crisis, increases in the price of petroleum caused depression and inflation in the member countries of this community, and affected their economy negatively for a significant time. Furthermore, although energy supply deduction only appeared in the short term (except in the First Gulf War), the world energy market was following a good course, and the EU energy supply became varied with the petroleum and natural gas coming from the North Sea.[184] The EU did not encounter any serious energy problems as a result of the

---

181 Energy, transport and environment indicators (2011), edition, Eurostat, European Commission.
182 **Krasner, Stephen (1983);** International regimes, Cornell University Press, from Krasners overview essay, p. 2.
183 **Erdogan, Murat (2011);** Middle East Policy of European Union, Report, Ankara, p. 12.
184 **Dieter, Helm (2005b);** European Energy Policy: Securing Supplies and Meeting the Challenge of Climate Change, New College, Oxford, 25 October. p. 2.

good relationships between the Central and Eastern European Countries that gained independence in the 1990s and the increase of performance provided in the use of energy. However, the EU was aware of its need for the development of an energy policy that can counter the risks related to the energy supply that it may encounter in the future.[185]

The problems concerning energy supply security threatened the EU, the biggest energy importer, especially as political problems between the EU and Russia heightened. The natural gas conflict between Russia and Ukraine in 2006 created a short-term energy supply crisis in Europe, affecting EU member countries and proving that exporter countries could use energy as a means of political compulsion. Russia wanted to increase the price of a thousand cubic meters of natural gas from 65 to 95 USD, and with this, the EU thought that Russia was using energy as a means of political compulsion against the western-oriented policy of the Ukrainian government. With these events, the EU began to question the reliability of Russia as an energy supplier. However, after the petroleum crises of the 1970s, the community, disregarding the USA's objections, cooperated with the Soviet Union to make the Soviet Union's natural gas available in Western Europe. After this process, the EU started to search for ways to vary the energy supply.[186]

Another crisis that affected the EU was the petroleum crisis between Estonia and Russia. Russia stopped delivering petroleum to Estonia by alleging that during the renovation of the railway of the Tallinn Port, the biggest port of the Baltic Sea, Russian petroleum was being transferred to the ships during the debates between Estonia and Russia about the removal of the Red Army monument in Tallinn, the capital city of Estonia. Russia decided to change its exporting port from Estonian to Russian harbours, the start of Russia's post-Putin energy strategy; giving priority to

Russian harbours instead of foreign ones.[187] In addition, the political problems surrounding Russia, issues such as the USA's attempt to establish the missile defense system in Eastern Europe to increase its defense against Iran, NATO's expansion towards the former Soviet Union countries like Georgia, and Ukraine's

---

185 **Pamir, Necdet (2005);** "AB'nin Enerji Sorunsalı ve Türkiye" (EU Energy Problems and Turkey), Stratejik Analiz Dergisi, Cilt.6, Sayı.67, p. 75.
186 **Westphal, Kirsten (2006);** Energy Policy between Multilateral Governance and Geopolitics: Whither Europe. Journal of International Politics of Society, p. 8.
187 **Tönjes, Christoph/Jong, Jacques (2007);** Perspective on Security of Supply in European Natural Gas Markets. Clingendael International Energy Programme Working Paper. August. p. 15–16.

approach to the EU, all damaged the Russia-EU relationship. Because Russia used energy as a political trump card and followed policies in this direction, the EU's energy dependence remained on the agenda.[188]

According to Tönjes, *"Russia was in search of using energy to receive support to its foreign policy demands from EU member countries. Russia makes the prices of energy different considering the political effect of the countries in political disagreements. Accordingly, Russia sold a thousand cubic meters of natural gas for 250 Dollar to Western Europe, 240 Dollar to the Baltic States, 230 Dollar to Ukraine, 235 Dollar to Georgia, 170 Dollar to Moldova, 110 Dollar to Armenia and 100 Dollar to Byelorussia. Even though Russia explains that these price differences are related to the transportation costs, it is known that Russia uses energy as a political device. Political opposition of the importer countries against Russia causes them to receive energy at higher prices."*[189]

It is supposed that demands for natural gas will increase in the next twenty years, especially in Europe and Asia, and with this increase in demand, competition for the natural gas reserves in the Caspian region is expected to occur between the EU, India and China. In addition to this, another subject forcing the EU to compete is the heavy increase of the share of liquefied natural gas (LNG) in the world's natural gas market. Natural gas supply's heading towards the world market in LNG form made Europe dependent on the natural gas prices in North America. This dependence is estimated to increase in the future.[190]

The roles of the international energy companies are important in providing the EU's energy supply security. Some European companies benefit from companies such as BP, Shell and Total in managing fossil fuels effectively. These companies take an important place in providing the security of the EU's energy supply. In recent years, the national petroleum companies of developing economies such as China, India and Turkey have come into prominence in the latest energy relations. Thus, new agreements have been made between the energy-nationalist exporters and the companies recently involved in the world's

---

188 **Copy, Van der Linde (2008);** Turning a weakness into a strength: a smart external energy policy for Europe. Clingendael International Energy, http://www.ifri.org/files/Energie/External4.pdf (accessed on 10/12/2012), p. 33.
189 **Tönjes, Christoph / Jong, Jacques (2007);** Perspective on Security of Supply in European Natural Gas Markets. Clingendael International Energy Programme Working Paper p. 16.
190 **Chevalier, Jean Marie (2005);** Security of Energy Supply for the European Union "European Review of Energy Markets", Vol. 1, Issue 3, November. p. 9.

energy market on the subject of energy commerce.[191] According to Linde, the USA's strategic plans have had an immense effect on the energy supply securities of EU countries in the Middle East and Asia, especially the USA's regime change in the Middle East. In spite of that, the military force that the USA used in the area improved the images of the energy importer countries like China. However, the USA's retreat from the area can pose a risk to the security of the energy flow. This situation has left the EU with a strategic dilemma about how the EU will provide its energy security in the future.[192]

Lastly, the Libyan crisis has caused a disruption of supplies. According to BP data, oil exports from Libya removed 1.2 million barrels per day of crude supply for the year. As another matter of subject, Japan's earthquake and tsunami disaster, which gave way to a nuclear crisis, affected other countries and other fuels. These events show that the world is interdependent, as any crisis can affect the entire world. The world energy consumption's damage to the energy markets is particularly sensitive to these kinds of crises.

5.6 Energy Targets and Prospects for the EU in 2020

EU presidents discussed agenda topics in the field of energy at the EU summit on 4 February 2011. The Commission specified five primary subjects and declared that a concrete legal initiative concerning these subjects would begin in 18 months. The EU Commission aims to focus on the biggest energy efficiency and potential in the transportation and building sectors. The Commission stipulated that the hosts should follow the required energy efficiency potential with the local financial support and investment initiatives, and that support in this subject should be carried out until the middle of 2011. Moreover, it was remarked that energy efficiency should be considered for service purchasing or selling in the public sphere, and that the proper technology for energy efficiency should be used in the field of industry. Another aim of the Commission is the completion of the EU's domestic energy market by the targeted date. All member countries are predicted to have ended this period by 2015 without any exceptions. The EU predicts that the energy substructure investment will need approximately one trillion EUR in the next ten years. Thus, the EU Commission states that the

---

191 **Keppler, Jan Horst (2007);** International Relations and Security of Energy Supply: Risks to Continuity and Geopolitical Risks, European Parliament's Committee on Foreign Affairs, Brussels, p. 5.
192 **Copy, Van der Linde (2008);** Turning a weakness into a strength: a smart external energy policy for Europe. Clingendael International Energy, p. 29.

EU should coordinate the strategy projects' structure licenses within the allotted time, which will accelerate the projects and bring them to completion. Another topic discussed is that there is no single voice of the 27 member countries in the world's arena to represent the field of energy. The Commission demanded that the EU coordinate the energy policy for relations with third-world countries. They encouraged promoting relations with neighbouring countries, and integrating these countries within the framework of the Energy Community Treaty in order for them to join the EU energy market. The Commission aimed to provide the necessity of energy for every citizen in the African continent, and declared that they cooperated mostly with this region. In addition to this, the EU is a leading advocate for the increase of investment in energy technologies. We can categorise the other subjects as supporting the new and smart technologies by providing a rivalry in the EU, backing up the energy, searching for secondary generation biofuels and increasing the energy efficiencies of smart cities. In the report, the Commission stated that there should be a rivalry related to the prices and between the distributors' companies in the field of energy, and that consumers should benefit from this. The EU's 2020 strategy aims for smart, sustainable and inclusive growth and looks to decrease greenhouse gas emissions by 20%, to increase renewable energy shares by 20%, and increase energy efficiency by 20% in concern of climate change.[193]

Summary

The European Union has set up its energy strategy and budget on a national basis for the coming ten years. With a population of 500 million in total, the EU is the second-biggest energy market in the world after the USA. Faced with its increasing energy consumption, the EU needs to both take precautions and undertake responsibilities. As it is mentioned in the study, the EU imports 50% of its energy and 64.2% of its gas. If the necessary investments are not made, it is estimated that, according to the EU Commission data, 65% of energy needs and 84% of natural gas needs will be imported from outside by the year 2030.

The share of gas in energy consumption was only 2% in 1960, however, the gas consumption in Europe has increased by 2% since 2000. According to the scenario, it is estimated that this increase will continue until 2030. Therefore, EU member countries will face high energy consumption and low gas and oil opportunities. Hence, they are required to diversify their natural gas import supply, or

---

193 **European Commission: 2011**, Energy 2020: A strategy for competitive, sustainable and secure energy.

seek alternative routes. At this point, according to Keohane and Nye, constructing alternative energy resources takes time and causes vulnerability for the importing countries in terms of prices. In the event of an interruption of the natural gas resources at an unforeseeable time, vulnerability is felt more. Until this period, Russia had transported the gas, which was transferred through Ukraine to EU member countries, however, they faced short gas interruptions during 2006 and 2009. Hence, the EU countries reviewed their relationships with countries such as Russia, Norway and Algeria, from which they imported their energy, and with Ukraine, which has the transit pipeline. Following these crises, the EU started to establish close relationships with the relevant countries in order to develop cooperation with the Caspian Sea and the Middle Eastern region in the field of energy. Initially the Nabucco project was developed, however, the project was suspended due to the lack of sufficient gas and financial shortcomings, and then it was narrowed down to West Nabucco. The TANAP project, launched as an Azerbaijani and Turkish initiative, plans to open the Southern Gas Corridor as TAP is selected as the exit route for the Shah Deniz Consortium. As opposed to this, Russia intends to continue its influence on the EU by activating the South and Nord Stream Pipelines. With the launch of these projects, it is estimated that EU gas dependence will increase and thus similar threads will be experienced. Meanwhile, Russia has great concern about the gas in relation to the transit countries. Both of the natural gas pipelines, the South and Nord Streams, will be transmitted directly under the Black Sea.

From this point of view, the member countries and the Commission have been striving on issues such as diversifying the gas supply dependency of the EU, increasing alternative energy resources and improving energy efficiency.

The EU could consider four regions as sources for gas imports:

1- North East European countries supply the gas from Norway first hand.
2- Central European countries import the gas mainly from Russia. The share of natural gas imported from Norway is low.
3- The majority of Eastern European countries are 100% dependent on Russian gas. With the South Stream and Nord Stream pipelines, the energy relationships between the countries will be boosted.
4- Some of the member countries supply gas from Egypt, Nigeria, Norway, Russia and Libya.

The EU's dependence on natural gas has direct and indirect effects. Since the EU is a supra-state organ, it should set up its strategies in accordance with the increasing consumption within the framework of EU energy policy. It is impossible to say that the energy policy of the EU is independent of the decisions taken at national level. The EU has the opportunity to mitigate foreign dependence through tailored projects and developed technologies. In this scope, support for energy chapter has been launched under the scope of EU research programmes and a support of 2.3 billion EUR was allocated for 2007–2013.

On the other hand, there are different energy policies between some members. Germany and Russia concluded the Nord Stream pipeline agreement in recent years, however, some member countries (Poland and the Baltic states) opposed this agreement. For EU countries, economic relations are important tools for competition. The ideal one is to have joint and single policies; however, the energy policy decisions are mostly in the hands of member countries. The European Commission desires that the member countries have a gain. New pipelines and pipeline routes reveal differences between the member countries. The EU strives towards setting common energy policies with regard to this issue. The European Parliament has prepared a report under the title of *Joint Energy Policy*.

5.7. Analysis of the Cooperation between the EU and Russia on Energy Policy

The largest gas reserve and the seventh largest oil reserve worldwide are in Russia, rendering Russia the biggest gas exporter and the second biggest oil exporter. The energy income of Russia more than meets the annual budget of the country. Oil and gas exportation equal 81% of the total exportation of the country.[194] The EU Commission mentioned relations with Russia in the 1995 report titled *The European Union and Russia: Prospective Relations*. In the report, it states that Russia should place emphasis on democracy and free market economy and that the level of political, economic and defense-based cooperations will be in parallel with the performance that Russia shows on the issues of democracy and liberalisation. The relations with Russia that started in 1992 continued with the Partnership and Cooperation Agreement that was signed at the Corfu Summit in 1994. Yet, this agreement not only includes the EU institutions but also the member states. At the end of the protracted negotiations, the agreement was put into action in 1996. After that, a mutual foreign and security policy was initiated with Russia in 1999, with the strategy document accepted by the EU, and for this purpose the Cooperation Council was founded. With the St. Petersburg Summit in 2003, the Cooperation Council was transformed into the Permanent Partnership Council.

---

194 Energy Information Administration: 2009, Russian Country Analysis Briefs. http://ei-01.eia.doe.gov/countries/cab.cfm?fips=RS (accessed on 09/11/2011).

The Partnership and Cooperation Agreement could not be extended due to the problems that Russia experienced with Poland and Lithuania. However, in 2008, the EU member countries helped renew the agreement by persuading the parties to not have problems in energy and commercial issues.

On the other hand, dialogue was initiated between the EU, Russia and the Eastern Bloc with the Lubbers' Plan, for the purpose of establishing cooperation in energy and supporting the EU energy security and the economies of these countries. As a result of this cooperation, the European Energy Charter was accepted in 1991 and put into action in 1998. Thus, by means of the EEC, the foundation of the relations between Russia and the EU was laid.[195] In 2000, after the energy charter between the EU and Russia, the necessity of founding an energy partnership was accentuated in this summit. The decisions taken were as follows: providing energy supply security in a legal framework; guaranteeing the physical security of the pipelines and networks that transport supply resources; establishing the required environment for foreign investments in the energy sector in Russia; assessing the primary projects that will increase energy savings; improving cooperation to increase security in the maritime and nuclear energy fields; raising environmental standards; an adaptation to the Kyoto Protocol; confirmation of the EEC; and ending negotiations on the Transportation Protocol. The Charter was signed based on these agreements.[196] Thus, the dialogue between the EU and Russia was settled but this dialogue could not reach the required level. The progress reports on the energy dialogue, which were published regularly, were limited to a narrow framework such as energy efficiency and energy savings. With the "Partnership and Cooperation Agreement" signed in Nice in 2001, the issues included in the predetermined energy partnership were more focused, and investments were made in energy sectors like oil, gas and electricity. As a result of the developing relations, Russia allowed European companies to buy shares in the Russian market and to contribute to the energy sector. The reports demonstrated that, in the improving relationships in energy, the aimed objectives were, to a remarkable extent, achieved and a lot of progress was made, which made a positive impact on the EU's energy security.[197]

---

195 European Commission: EU-Russia Energy Dialogue http://ec.europa.eu/energy/international/russia/dialogue/dialogue_en.htm (accessed on 09/11/2011).
196 **Aras, B./Yorkan, Arzu (2005)**; Avrupa Birliği ve Enerji Güvenliği:(EU and Energy Security) Siyaset, Ekonomi ve Çevre (TASAM Stratejik Rapor), No:13, Ankara, Tasam Yayınları., p. 7.
197 European Commission: EU-Russia Energy Dialogue http://ec.europa.eu/energy/international/russia/dialogue/dialogue_en.htm (accessed on 09/11/2011).

Figure 10: Russian Exports

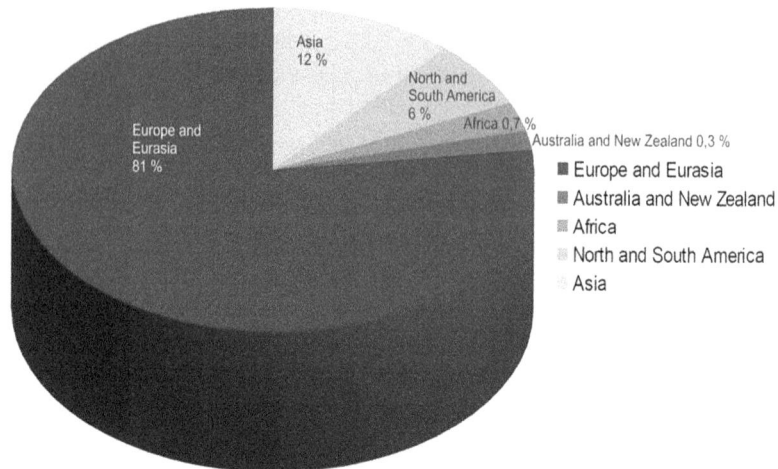

Source: Global Trade Atlas, Facts and BA

As seen in the figure above, 81% of Russian oil exports go to Europe and Eurasia, while 12% go to Asia and 6% go to North and South America. More than half of Russian oil exports, 50% in oil and 64% in gas, go to the EU. Russia wants to increase this export rate and aims to improve relations with the EU, who is its most important trade partner in the field of energy. Russia's targeted total energy export to the EU in 2011 was 530 MTEP and this is expected to reach 565 MTEP by 2020. The expansion of the EU into Eastern and Central Europe has led to a stronger emphasis on Russia Eastern and Central European countries regard Russian policies as a threat to the energy supply security. For instance, the Russian gas cut, which was applied by Ukraine and which influenced the whole world, ended with a pricing disagreement between two countries. The underlying reason for the crisis was Ukraine's effort for inclusion in NATO and the EU pulling away from Russia's influence. After the gas crisis between Ukraine and Russia, the anti-Russian Timoshenko government was subverted and a pro-Moscow government led by Yanukovych was established. The new government declared that the increased gas rate was accepted. After this process, Ukraine became the country that paid the cheapest price to the Commonwealth of Independent States among member countries. The 2008 Russian operation in Georgia, Abkhazia and South Ossetia was not only a military movement, but it was also seen as a move that aimed to prevent the western market from accessing oil and gas in the region, through a route avoiding Russia.[198]

On the other hand, the power crisis that arose from petroleum taxation in Russia during 2007 was the second major power crisis affecting Europe. Russia had not been levying the petroleum exporting tax from Belarus enfranchised by private arrangements. Afterwards, Russia called to retract the enfranchisement of Belarus, on the grounds that Belarus was exporting and levying the petroleum that it had been importing from Russia without tax. From thereon, Belarus has required the transit tax from Russia. During the conflict, an interruption in Europe ensued for three days.[199] In this context, the EU has become concerned about similar crises occurring, due to its increasing power demand and its power dependency on Russia. It is supposed that this dependency will increase to 50% for fuel and 80% for natural gas imports from Russia by 2030. The power dependency of member states on Russia is not equivalent.[200] Russia is accustomed to a gain in political and economical profit by using energy as an oppressive tool against EU countries. In this regard, the power strategy of Russia has been to try and make Europe more dependent on Russian fuel and natural gas. Andrew Monagham ascertains the Russia-EU correlation: *"Putin sees power as an important political power and this power develops the economical power of Russia as well. Some experts think that Russia wants to control Europe's social overhead capital and make its power security dependant to itself. In this perspective, Russia would create the most prolonged and integrated market of the world if the EU establishes only one successful market in the field of electricity and gas."*[201] Meanwhile, the analyst Paul Belkin claims there is an ambiguity in EU power politics: '' the EU Comission and the EU countries sometimes differ from each other on energy security policy. Russia and Germany have made a long-term agreement for Baltic pipelie. On other hand, LNG gas agreement is signed by Spain, France and Algeria. The member states decide about their energy strategy by analyzing national security. Those nationals long- agreements can be theaten to the EU . By no means, it is highlighted that it needs an extremen coordination to execute this policy by the EU-

---

198 **Woehrel, Seven (2008);** Russian Energy Policy Toward Neighboring Countries. CRS, Report for Congress, November 27, p. 7.
199 **Stern, Jonathan (2006);** Security of European Natural Gas Supplies: The Impact of Import Dependence and Liberalization, The Royal institute of International Affairs, London, pp. 3–24.
200 Energy Information Administration: 2009, Russian Country Analysis Brief. http://ei-01.eia.doe.gov/countries/cab.cfm?fips=RS (accessed on 09/11/2011).
201 **Andrew, Monagham (2006);** Russia-EU Relations: An Emerging Energy Security Dilemma, Pro et Contra, vol. 10, issue 2–3 (Summer 2006), Carnegie Moscow Center.

Comission"[202] Consequentially, Russia prefers to make agreements with EU member states individually but not gregariously. In this way, Russia discriminates the price among the states' importing power via these agreements. Hence, Russia requires the highest price that each state can manage.[203] As an example, Russia made an agreement with Greece and Bulgaria for a petroleum pipeline construction in 2007. This "Burgaz-Dedeağaç Petroleum Pipeline" would be the first petroleum pipeline under the control of Russia within the borders of EU. On the other hand, Russia decided to cooperate with Gazprom to expand Turkey's Blue Stream Pipeline in Hungary. The Russian and Italian company ENI have signed a memorandum of understanding to construct the South Stream Pipeline from Russia to Italy, ensuring a transition to the Gazprom Italy distribution system through the Italian company ENI.

5.7.1 Russian Gas Pipelines and Projects to Europe

After the United States, Russia is the second largest producer of natural gas in the world. The EU's dependency on imports of natural gas reached 64.2% in 2010, compared with 48.9% in 2000. Sixty-four percent of imported natural gas came from Russia, followed by Norway and Algeria.

Figure 11: Share of Russia's natural gas exports by destination, 2010

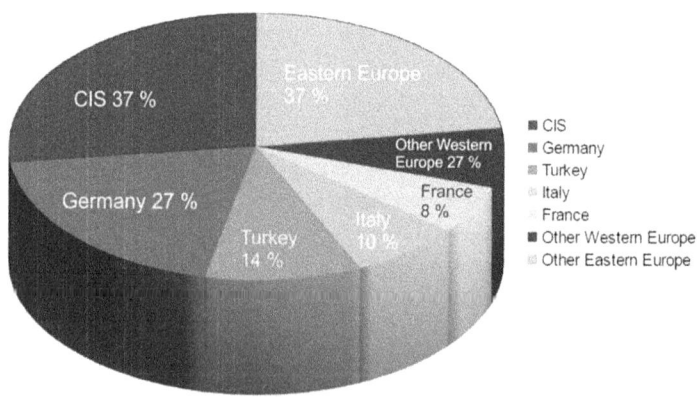

Source: Eastern Block Energy, US Energy Information Administration

---

202 **Belkin, Paul (2008);** Report for Congress: The European Union's Energy Security Challenges Updated January 30, Analyst in European Affairs Foreign Affairs, Defense, and Trade Division, p. 23.
203 **Cohen, Ariel (2007);** Europe's Strategic Dependence on Russian Energy, p. 3.

According to this data, Russia is the main gas supplier of the EU. Russia wants to increase this dependency with new projects such as the South Stream gas pipeline, for which construction has already started. This pipeline will transport Russian gas to 38 million households in southern Europe in 2015. The pipeline length is 2,400 km, traversing the Black Sea and carrying Russian gas via Serbia, Hungary, and Slovenia to Italy. *"Turkey allowed construction of the South Stream gas pipeline, and Russian access to Turkish waters of the Black Sea allowing Russia to bypass a hostile Ukraine to Southern and Central Europe. The total capacity of the South Stream aims to diversify Russia's energy routes to European markets and will, per year, deliver 63 bcm to Europe."*[204]

5.7.2 South Stream

The South Stream pipeline project was signed by Austria, Bulgaria, Croatia, Greece, Hungary, Serbia and Slovenia between 2008 and 2010. This pipeline project aims to span from the Russian coast to the Bulgarian coast by spreading the pipeline from Russia through the Black Sea. This pipeline is supposed to be over 925 km under the surface of the Black Sea. It is estimated that 63 billion cubic meters of natural gas will be carried via this pipeline. The EU aims to secure their power supply through the South Stream Project. Thus, the EU has begun a strategic project that diversifies the power supply in the gas field. The South Stream pipeline presents a direct power supply between the vendor and user to EU continental states. Analysts believe that the EU's gas demand will increase in the middle and long term. Increasing gas consumption by EU states in the industrial area and their preference for eco-friendly natural gas instead of coal, fossil fuel and nuclear energy also increase the demand in the marketplace. Hereafter, it is estimated that the EU will need to import more gas. According to the studies, it is estimated that the EU's gas export will increase to around 80 billion cubic meters in 2020, and these rates will increase to around 140 billion cubic meters in 2030. According to the acquired data, the construction of the South Stream pipeline has begun, and therefore the EU will export gas from Russia without any risk to the EU's power supply security.[205] In 2011, the project gained acceleration with Turkey's approval of the project passing through its borders. In return, Turkey received a discount from Russia for gas valued at one billion dollars. In this sense, the Energy Minister of Turkey, Yildiz said *"Although in the short term it might look like rival projects, in the mid and long term they are not. As Europe needs*

---

204 European Commission: Eurostat (2012); http://epp.eurostat.ec.europa.eu/statistics_explained/index.php/Energy_production_and_imports (accessed on 10/13/2013).
205 South stream official page: Europe's Energy Security http://south-stream.info/index.php?id=9&L=1 (accessed on 12/10/2012).

more natural gas, there will be the need for an additional three or four projects like this. This is why we have a positive approach to South Stream." He added that *"Turkey has no intention of becoming a shareholder in the South Stream project; however, Turkey plays an important role in the TANAP and TAP projects."*[206]

Figure 12: Map of South Stream Pipeline Route Option

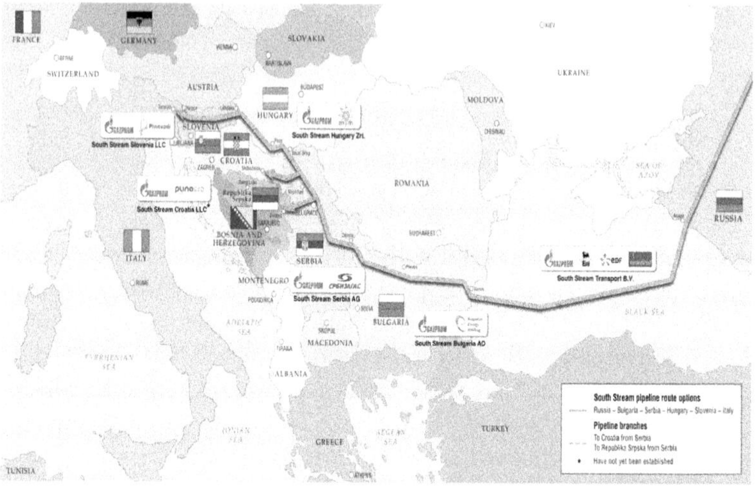

Source: Gazprom

There are several arguments concerning this project. Many experts assume that the South Stream Project aims to monopolise control of the transport of petroleum and natural gas from the Caspian region to Europe. Furthermore, this pipeline will provide uninterrupted delivery of gas to Europe and has been designed to bypass Ukraine, a country that has had political conflicts with Russia in the past. On the other hand, Gazprom authorities indicate that the initial cost of the South Stream natural gas pipeline project—20 billion USD—has now reached to 40 billion USD. This increase in the cost of the project is estimated to cause a burden on the states. Gazprom, which attaches the increase in costs mainly to the environmental conditions, is concerned about this increase of costs in the construction of Gazprom South Stream.[207]

---

206 http://www.hurriyetdailynews.com/russia-starts-16-bln-euro-gas project.aspx?pageID=238&nID=36403&NewsCatID=348 (accessed on 05/04/2013).
207 http://enerjienstitusu.com/2013/02/14/guney-akimin-akibeti-ve-gazpromun-foyasi/ (accessed on 03/07/2013).

5.7.3 Nord Stream Pipeline

The Nord Stream pipeline is a new direction for Russian gas exports to Europe, and aims to supply gas to Germany, the UK, the Netherlands, France and Denmark. This pipeline will be constructed by Gazprom and the German partnership companies EON and BASF. The Nord Stream has two parallel lines, the first of which was laid in May 2011. The second line was laid on 8 October 2011, and is the longest sub-sea pipeline in the world. It is claimed that the pipeline may cause a conflict in the EU's power security if it does not settle a consensus with Russia. In this context, several member states have expanded their dependency by signing a mutual agreement with Russia. Germany and Italy are the supreme exporters of Russian gas in particular.[208]

Figure 13: Map of Nord Stream Pipeline Route Option

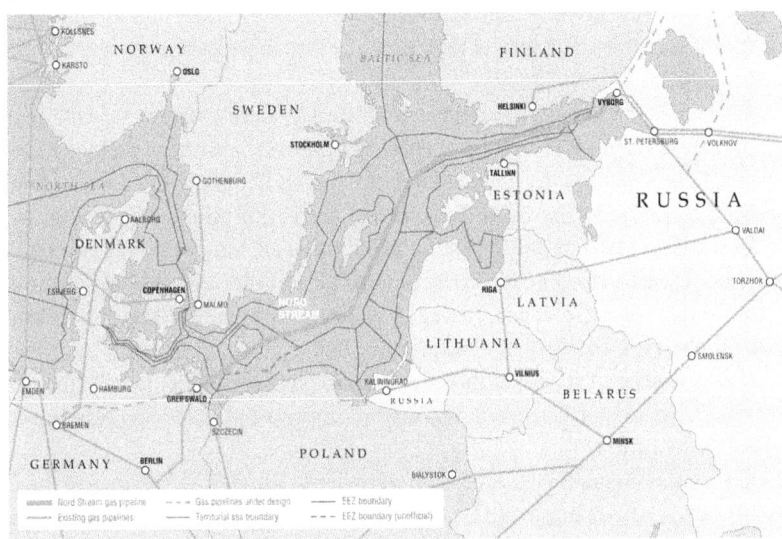

Source: Gazprom

It is predicted that EU gas imports will increase in the coming decades. Nord Stream is a significant project that does not require transit countries, thus reducing its gas transmission costs. The proposal project will connect Russia's

---

208 **William, Drozdia (2006)**; Russia: More Awkward, But Still Indispensable, European Affairs, Vol. 7 Issue 1–2 in the Spring/Summer of 2006.

Baltic Sea coast near Vyborg with Germany's Baltic Sea coast in the vicinity of Greifswald. The length of the project is 1224 km. Moreover, Nord Stream has additional transportation capacity with a third pipeline. With the Nord Stream project, Russia demonstrates that it has broken its dependence on the Baltic countries.[209]

5.7.4 Gazprom within the State

The Soviet Gas Ministry carried out its activities in gas production and distribution in the Soviet Union period in 1965. In 1989, Gazprom continued its activities as a company under state control with the title "Russian Gas Ministry". The company was fully under state control until privatisation started between 1993-1997. Thus, it initially engaged in the extraction of sources such as gas, oil and hydrocarbon and marketing.[210]

Russia's biggest investor abroad is Gazprom. Most of these investments are in Europe. One of the main aims of this company is to increase the gas market share in Europe from 27% to 33%.[211]

Gazprom is mainly engaged in gas trade and it operates in the transit, distribution and storage sectors. It has expansion plans in countries such as Russia, Austria, Belgium, the UK, Italy, Poland, the Czech Republic, France, Germany, the Middle East and the USA. Despite the increasing monopoly of Russia in the EU, Russia restricts the investing capabilities of European gas companies in Russia. Despite this, the EU governments also hinder investments in the gas sector. In this regard, Gazprom tried to acquire the biggest UK gas distributor in 2006, however, the British Competition Board changed its rules in order to block Gazprom and its entry into the UK market was prevented. It thus prevented the actions of Gazprom and any attempt to politicise any possible gas supply in the future.[212]

On the other hand, the USA has concerns about Russia's dominance in the EU gas market. As an example of this, the US Secretary of State, Condoleezza Rice threatened the Greek government regarding the participation of Gazprom in the

---

209 Gazprom official webpage: http://gazprom.com/about/production/projects/pipelines/nord-stream/ (accessed on 05/04/2013).
210 http://www.rferl.org/content/article/1064448.html (accessed on 03/07/2013).
211 http://www.gazprom.com/about/marketing/europe/ (accessed on 03/07/2013).
212 **Andrei Milovzorov**; "Gazprom navodit uzhas na Evropu", Ekonomika, March 3, 2006.

Turkish–Greek Interconnector project in the April of 2006. The US desires that the Turkey and Greece's trust in Russian gas decreases.[213]

The expansion of Gazprom in Europe increases the dependence between the EU and Russia. However, this expansion creates concern among the European countries, who impose limits in order to protect the gas market.

Summary

The EU and Russia have interdependence in economic relations. Both parties strive towards having a more advantageous condition against the other. The way to overcome this interdependence could be the diversification of the supply and demand between the commercial partners. In this process, the EU is more dependent on Russia. The EU is dependent on Russian gas at 64%. Moreover, this rate reaches 100% in some Eastern Europe countries. In addition to this, some Western European countries need the Russian gas. EU countries are required to pay a certain cost for diversification of energy supply; it is necessary to make investments in constructing new pipelines and establishing the infrastructure.

On the other hand, Russia strives towards diversifying the import supply. The discovery of new gas areas and the development of new liquid gas technologies in Eastern Russia requires a high amount of investment to be made for Gazprom in the coming years. Meanwhile, finding new resources will create an advantage for the domestic market. On the other hand, selling the gas for a cheaper price in the domestic market also brings a competitive advantage in the field of industry in the country's energy consumption.

The energy needs of China have been increasing rapidly, causing an increase in the prices of oil and coal. The problem of interruption of the gas supply between Russia and Belarus and Ukraine has had negative impacts on Russia. Following this process, the European Commission took the joint decision on the diversification of energy supply. It increased its support first on Nabucco and then on the TANAP and TAP projects.

On the other hand, some countries supported the South and Nord Stream projects, while others opposed them. As a consequence, it was understood that the EU did not make a joint decision on energy policy.

Besides, another important question that should be asked is whether the interdependence between the EU and Russia is symmetrical. Which of the parties

---

213 Kerin Hope, "Rice Bid to Exclude Russian Supplier", Financial Times, April 25, 2006.

could implement the strategy of energy supply diversification and become more advantageous economically compared to the other party? Can the rate of interdependence that occurs be considered as fast, or does the alternative option increase the cost? According to Keohane and Nye, the actors should be damaged less in interdependence. Russia is vulnerable here as the exporter country and the EU is at an advantage as the importer country.

Russia wishes to increase the dependence of the EU with Nord and South Stream pipelines. Opposed to this, the EU seeks new opportunities and intends to decrease its dependence on Russia by sourcing its gas from Azerbaijan with the TANAP project. How can sensitivity be applied to changes between the actors and what is the cost of its harmonisation?

The sensitivity relationship is considered in the relations between the EU and Russia that occur in short- and long-term alternative capacities. With the interruptions of supply for four days in 2006 and eleven days in 2009, EU member states recognised their dependence on Russia as the supplier country as the impacts of these interruptions were felt by 18 EU member countries.[214]

These countries did not have sufficient gas capacity during the supply interruption. Besides, the EU could not fulfill its planned target. Following the crisis, an abstract was published in an EU press release in relation to the crisis management of the countries that were affected. Germany supplied gas to Slovakia through the Czech Republic, and to Hungary, Slovenia and Balkan countries through Austria and the UK. It increased its gas production in that process and exported gas. Following the official statement of the EU, the governments and the non-governmental European Commission tested their operational ability. Meanwhile, it was also thought that the EU did not respond sufficiently to the gas interruption.

When assessed in political terms, following the crisis between Russia and Ukraine, it was seen that the energy supply was used as a threat to the safety of EU member countries. As a consequence, the supply interruption that developed instantaneously is considered to have had a high sensitivity impact among the member countries. Mitigating the dependence on imports is important in terms of the diversification of long-term energy supply. In this regard, the TANAP and TAP projects are supported by the Commission. Besides,

---

214 (http://diepresse.com/home/wirtschaft/economist/444884/index.do?_vl_backlink=/home/wirtschaft/economist/441630/index.do&direct=441630; 12.09.2010 und http://www.europarl. europa.eu/sides/getDoc.do?pubRef=-//EP//TEXT+IMPRESS+20090202STO47914+0+DOC+XML+V0//DE; 12.09.2010).

this step will decrease the sensitivity against Russia. However, the Nabucco project was abandoned and alternatively the TANAP project was developed. As opposed to this, Russia and Gazprom acted rapidly and in a target-oriented manner to a wider extent. Within a short period of time, negotiations in the political arena were launched with Ukraine at a senior level. Vladimir Putin and the President of Ukraine, Julia Timoschenko, met in Moscow and the parties agreed on the price of gas. The political crisis was resolved without the pressure of the EU.[215]

Similarly, with the construction of the Russian South and Nord Stream projects, the dependence on exports is intended to increase. Russia and Gazprom act rapidly with these projects, increasing the sensitivity against the EU.

5.8 Alternative Policy Options for Reducing Gas Dependency

5.8.1 Promoting Use of EU Renewable Energy Sources

While the world currently meets 2.5% of their energy demands from renewable energy resources, according to the IEA, this demand will be 3.3% in 2015. According to the data of the same agency, an investment of approximately 10.5 trillion EUR will be made in renewable energy resources in the period between 2001 and 2030. It is expected that the rate of renewable energy resources in the production of energy will reach 25% in OECD countries.[216] Member countries determined goals for the future of renewable energy to deal with the problems related to the security of energy supply and climate change, in accordance with the advice from the EU Commission at the March 2007 summit. In this context, they aim to increase the renewable energy rate to 20% of the EU's total energy consumption, and to increase the usage of biofuel in the transportation sector to 10% by 2020. However, according to data from the International Energy Agency, renewable energy consumption is predicted to be at the level of 14% in 2020.

---

215 http://diepresse.com/home/wirtschaft/economist/444349/index.do?_vl_backlink=/home/wirtschaft/economist/441630/index.do&direct=441630; (Accessed 12.09.2010).
216 World Energy Council, Turkish National Committee, Energy Report, Ankara, 2010, p. 85.

Figure 14: Share of renewable energy in gross final energy consumption and target for 2020

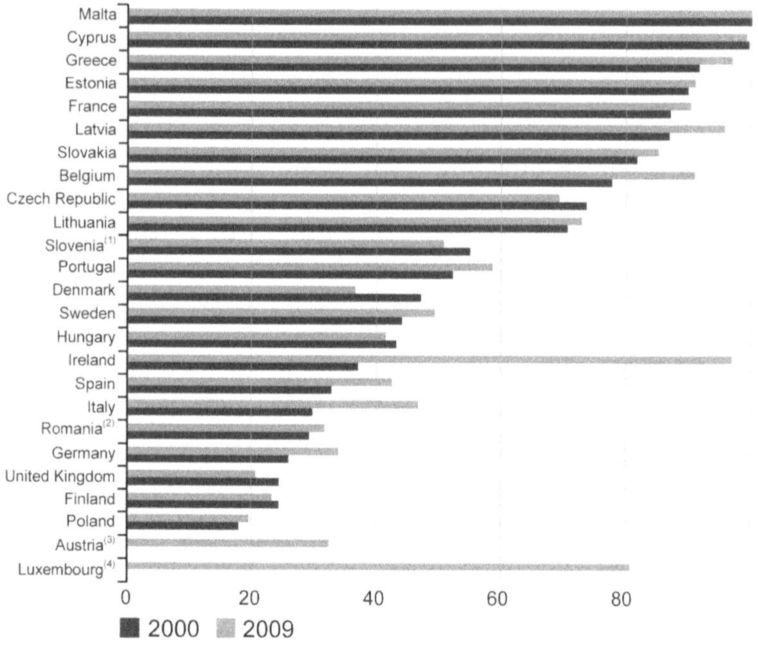

[Bar chart showing renewable energy shares for EU countries in 2000 and 2009, with countries listed from Malta (top) to Luxembourg (bottom): Malta, Cyprus, Greece, Estonia, France, Latvia, Slovakia, Belgium, Czech Republic, Lithuania, Slovenia⁽¹⁾, Portugal, Denmark, Sweden, Hungary, Ireland, Spain, Italy, Romania⁽²⁾, Germany, United Kingdom, Finland, Poland, Austria⁽³⁾, Luxembourg⁽⁴⁾. X-axis from 0 to 80, with legend showing 2000 and 2009.]

Source: Eurostat (Europe 2020 indicators)

The EU's aims for 2020 energy production can be seen in the above table. The EU, which has 27 members, aims to provide 20% of its total energy production from renewable energy resources. According to the Commission's report in 2009, the total renewable energy share was 10.3% in 2008. Sweden is the leading country that held the biggest share of renewable energy in total, with its production of 44.4% in 2008. Finland (30.5%), Latvia (29.9%), Austria and Portugal (28.5%) follow Sweden. Luxembourg (2.1%) and England (2.2%) are at the bottom of renewable energy production. Austria became the country that developed its renewable energy production the most by increasing it to 28.5% in 2008, up from 24.8% in 2006.[217] On the other hand, when the electricity production from renewable energy is considered, the EU gradually achieved growth at the rate of 6.4% between 1999 and 2009. When energy sectors are examined, it is seen that the

---

217 Energy, transport and environment indicators, 2011 edition, Eurostat, European Commission, p. 71.

greatest growth related to renewable energy is in wind energy. When electricity production from wind energy is considered, the share of 6% in 1999 increased to 29% in 2009.

Figure 15: Installed capacity for electricity generation from renewables, EU-27

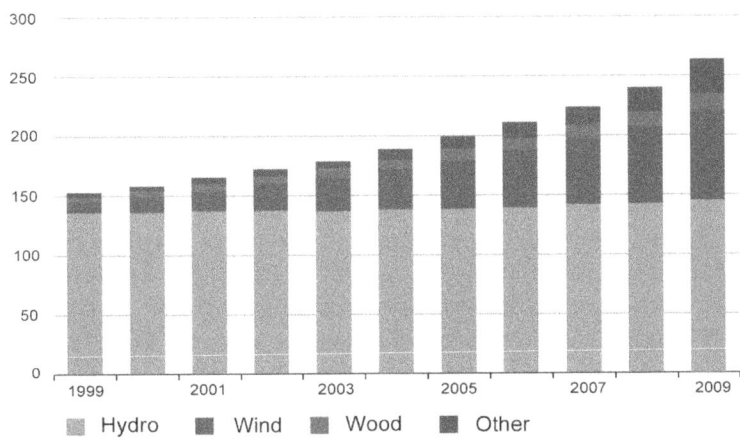

Source: Eurostat

The other sectors given in the table relating to renewable energy are thermal photovoltaic and biogas, and their shares increased by eight from 1999 to 2009. In these sectors, the rate that was 2% in 1999 increased to 11% in 2009. Hydroelectric has the highest rate of energy production from renewable energy with a rate of 55% in 2009. Electricity production from renewable energy in the EU was 587 Twh in 2009. When this is compared to the production in 1999, it can be seen that it increased by 50%. When the EU's total electricity production is considered, production rose to 18.2% in 2009 from 13.2% in 1999. When the share in the member countries is considered, Germany produced 95 Twh, Sweden produced 80 Twh and Spain produced 74 Twh of electricity.[218]

5.8.2 Improving Energy Efficiency

EU energy and environment policy is formed at the same time as energy efficiency. One of the important goals of the Kyoto Protocol is energy efficiency. The

---

218 Energy, transport and environment indicators, 2011 edition, Eurostat, European Commission, P. 73.

EU Commission drafted a plan on rationalistic usage of energy and providing energy efficiency in April 1998. The Commission determined its movements and activities in this subject under the title of the EU SAVE Programme (smart energy for the EU) for 2010. Thus, the energy efficiency policy and coordination between the member countries started. The SAVE programme came into effect under the directive of energy efficiency 2006/32/EC.69. A budget of approximately 66 million EUR was reserved for this programme between the years 1998 and 2002.[219]

The IEA suggested some applications in the direction of applying the twelve primary precautions globally in 2007. Firstly, energy efficiency would be applied around the world as an effective means of achieving climate change goals. Thus, it was declared that they would save energy at the rate of net 9% between the years 2008 and 2016 with energy efficiency projects that started in the 1970s. The USA and some other developed countries hesitated to declare their carbon emission status; however, they participated in this process more effectively with the Copenhagen Climate Summit. Some countries, especially EU countries and Japan, declared their goals for carbon emissions and energy saving; with this declaration, it was expected that more investments in efficiency for industries would be made.[220] Furthermore, changes in industries and energy efficiency in the building and transportation sector will play a significant role in decreasing greenhouse gas emissions, especially within the building sector, which constitutes 40% of the EU's total energy consumption. The EU Commission aims for a 20% energy saving, which should be possible as long as all necessary applications are carried out by 2020.

Consequently, EU countries reached an agreement to decrease greenhouse gas emissions by 2020 when compared to the year 1990. However, when the EU became party to the international Kyoto Protocol, this rate rose to 30% for 2020. In addition to this, it was aimed to ensure that the fuel used in the transportation sector be comprised of biofuel at a rate of 10% in 2020. Jose Manuel Barrosso, head of the EU Commission, predicted that 87.7 billion USD would be saved as a result of these applications and goals.

---

219 Directive 2006/32/EC of the European Parliament and of the Council of 5 April 2006, pp. 64–85.
220 Türkiye Enerji ve Enerji Verimliliği Çalışmaları Raporu (Report of Turkey Energy Studies and Energy Efficiency) "Yeşil Ekonomiye Geçiş" Temmuz – 2010.

5.8.3 Nuclear Energy Use and Debate

Nuclear energy is as important for the EU as coal, and is discussed in this section because of its possible damage to the environment. There are 176 nuclear power plants in the EU, and the share of nuclear energy for energy consumption in the EU is 13.4%, according to 2007 data. The Union has not yet made a net decision relating to nuclear energy, and currently the EU Commission allows member countries to make their own decisions on the subject of nuclear energy usage.

Figure 16: Share of Nuclear in National Electricity Generation in 2009

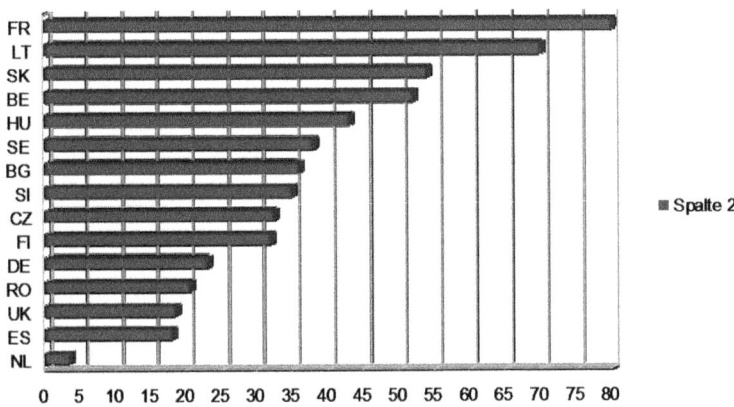

Source: Market Observatory for Energy 2011, Eurostat May

The rate of electricity produced from the nuclear power plants of fifteen member countries in 2009 is given in the above table. According to this table, France has the highest rate of nuclear energy production at 76%. When nuclear energy developments at the time are considered, one power plant met the 1600 megawatt energy consumption of a city with an approximate population of 1.5 million in the north of France in 2014. Sweden and Germany aim to gradually close all nuclear power plants. Austria, Finland and Luxembourg are EU countries that do not have any nuclear plants. All the nuclear energy produced in the EU is used in the production of electricity energy. Thus, EU countries meet 35% of their electricity needs from nuclear power. Another point underlying EU Commission's ending of its objective approach to nuclear energy is the desire to decrease energy dependence on Russia, especially member countries' dependence, and to increase the attention paid to nuclear power plants. The EU Commission is considering building nuclear power stations in these countries, and wants the assignation of the arrangements related to the standards of nuclear energy plants to fall within its own area of jurisdiction.[221] In the green paper the EU declared that as a result of environmental problems the main future policy change within the energy field will be to decrease nuclear energy usage. An increase of consciousness regarding the risks of nuclear energy resulted in the early closure of power plants in 1990, and an increase in security precautions after the Chernobyl accident and Fukushima disaster has taken place. Owing to the increase of energy prices and the improvement in the security precautions of nuclear plants in the EU, some member countries have started to reevaluate their positions on nuclear power usage. Especially in 2010, important developments were seen in nuclear energy, and changes in the policies of many countries have since been made. For example, Sweden began to revise its decision to close nuclear power plants gradually through a nuclear moratorium with a referendum. Another EU country, Italy, is performing research in the direction of gaining control of nuclear energy again. The different methods for change show that decisions regarding the future of nuclear energy vary amongst the EU's member countries, and that there is still much potential for change in this field.[222]

5.8.4 Single Energy Market and Liberalisation of the Gas Market in the European Union

There were various considerations concerning the integration of energy markets and the liberalisation of the domestic energy market by providing a competitive environment in the 1980s, and in this context, a white paper titled *Energy Policy* was published for the EU. As mentioned, this paper was published to establish the general principles and aims for the EU's domestic energy market in 1990. In the frame of a common market, a procedure was initiated for competition with the USA in order to liberalise the energy sector, and thus to provide liberal energy trade based on complete competition. In this context, the EU took steps to constitute the competitive environment in the domestic energy market by breaking down the monopoly position that national energy foundations had.

The first component of EU energy policy was to create a domestic energy market. Firstly, the base of creation of the domestic market was competitiveness, and with this, the EU domestic energy market began to be shaped with the Single European Act. The coal market was the first of the three founding agreements

---

221 **European Comission (2011);** Market Observatory for Energy.
222 **World Energy Council, Turkisch National Committee (2010);** Energy Report, Ankara, 2010, p. 8.

in the first foundation years; the nuclear energy market and the domestic petrol market that joined the community quickly in the 1970s and still has strategic importance; the domestic natural gas and electricity market composed of natural gas whose importance was appreciated all around the world in the 1990s; and lastly, the market of electricity, a constant energy element, constituted the EU domestic energy market.[223] Pursuant to the Single European Act, the liberalisation procedure in the sectors of telecommunication and electricity supply was the aim of 1987, both because electricity and telecommunication sectors were in the hands of national monopolies. However, because the limits of the electricity sector are more protected than those of the telecommunication sector, liberalisation in the telecommunication sector occurred first. However, countries such as France were not willing to allow liberalisation in the electricity sector, and this opposition extended the period.[224]

The Common Electricity Market Directive that the Council determined in 1996 was put into force on 1 February 1997. With this directive, some targets were set regarding the liberalisation of both the production and distribution of electricity. Thus, the liberalisation of electricity markets of all countries was aimed to be at least 25.3% in 1999, and this rate aimed to reach 28% in 2000, 32% in 2003, and for all markets to be liberalised by 2007. The directive of natural gas and electricity in force predicted a two-stage liberalisation: firstly, all industrial, commercial and professional users (except for domestic consumers) were able to choose their own suppliers, and secondly, the domestic consumer could choose their own suppliers as of 1 July 2007.[225]

Even if the liberalisation of EU electricity and natural gas markets is founded with the directives of electricity and natural gas, it is thought that some steps for constituting the EU's completely liberalised energy market should be taken. The period of market liberalisation started to appear in Europe as a result of electricity and natural gas directives. Priorities in the field of the liberalisation of energy markets were not the same in all countries. While France, England and Italy

---

223 **Andersen, Sevin S. (2000);** European Integration and the Changing Paradigm of Energy Policy: The Case of Natural Gas Liberalisation, Arena Working Papers, pp. 6–10.
224 **Ian, Bartle (1999);** "Transnational Interests in the European Union: Globalization and Changing Organization in Telecommunications and Electricity", Journal of Common Market Studies, vol. 37, issue 3 pp. 367–374.
225 Directive 2003/55/EC of the European Parliament and of the Council of 26 June 2003 concerning common rules for the internal market in the natural gas and repealing Directive 98/30/EC.

were in a national structure, Germany and Holland came into prominence with regional settlement.

A decrease in the prices was expected; firstly through competition related to the usage of effective management and investment techniques in the frame of EU competition rules in the energy markets, and secondly by putting pressure on prices and activity. The power of competition increased in the world markets of the European industry as a result of the competition in the electricity and natural gas markets. With the increasing activity, lower costs and prices emerged, and thus the power of competition in the industry increased. It is known that the price factor is important in the international arena. For the reason of enhancing social welfare and increasing the power of competition in industry with the competitive energy system, low energy prices for steel, paper and motor vehicles—for which energy is essential—are important in terms of the power of competition.[226]

The result of this increased competition and liberalisation meant that the quality of the services presented in the field of energy started to increase; a result of consumer choice, as firms with poor service qualities risked losing their customers. The companies had to make investment in new technologies to increase their efficiency by applying different instructions for customers to be able to compete. The tendency of making innovations in the liberalised market offers different alternatives for the customers in the field of purchasing energy.

In addition to the improvement of quality, the liberalisation of the EU energy markets influenced environmental concerns positively. Opportunities for the customers to favour electricity produced from environmentally friendly energy resources in the member countries were results of the liberalisation. In addition to this, firms indicated payment alternatives, making customers them contribute to a fund that supports the investments for protecting the environment by offering different alternatives to customers in the liberalised market.[227] Thus, energy companies were inclined towards environmentally friendly production to gain competitive power. Especially after the liberalisation of the natural gas market, the price of natural gas decreased, and as a result, usage of natural gas in the production of electricity became more common.

In other words, the advantages of environmentally friendly energy production in terms of the power of competition encourage customers to lean more towards

---

226 **Tonus, Özgür (2004);** "Genişleyen AB'nin Enerji Politikaları ve Türkiye", (EU Energy Policies and Turkey on Process of Negotiation) Müzakere Sürecinde Türkiye AB İlişkileri Uluslararası Sempozyumu, Ankara, Gazi Üniversitesi, p. 9.
227 **Ergun/Çağdaş Evrim (2007);** Avrupa Birliği Enerji Hukuku (EU Energy Law), Caköak yayinevi Ankara, p. 101.

environmentally friendly energy resources and to use environmentally friendly technologies. Especially after the liberalisation of the natural gas market, decreases in the prices of natural gas and increases in the usage of natural gas in the production of electricity positively affected the protection of the environment. The rate of natural gas usage in the EU for the production of electricity between 1990 and 1998 increased by 128%, and the rate of solid fuel usage decreased at a rate of 15% in the same period. While the rate of natural gas usage in the production of electricity in the EU in 1990 was 11%, this rate became 24% in 1998, and the rate of carbon concentration decreased at a rate of 15.2%. For example, the rate of carbon dioxide emissions in England decreased at a rate of 28.5% between 1990 and 1999.

In the event that a common energy policy is constituted concerning the development and application of EU energy policy, an improvement can be the point in question. In this sense, EU member countries need to create a common market in the field of energy and assign their national decisions to the EU. Thus, consumers can make use of price reductions by providing competition. The EU needs to liberalise the domestic markets of the nations in the field of energy to create a common market. Germany can have an important effect, in terms of both its economic and political significance.[228]

---

228 Bundesministerium für Wirtschaft und Technologie (2010); Energie in Deutschland, p. 30.

6. Turkey's Energy Dependency and Role in the Southern Gas Corridor

With an annual growth rate of 1.7%, Turkey's population is expected to increase from 72 million in 2011 to 83.4 million in 2022. Along with population growth, Turkey's GDP increased by approximately 6.9% annually between 2002 and 2007, and is forecasted to rise 6.4% annually until 2020. As a result of its growing population and expanding economy, Turkey's total final energy consumption is projected to rise by 5.9% annually between 2011 and 2020. Turkey has been one of the fastest growing countries in the world, both in terms of economic and TFC growth in the last decade. Turkey's usage of energy has increased by leaps and bounds since the 1970s. Turkey's generation of primary energy sources was abundant and low-cost, giving Turkey high power supply security, which also explains the increase in energy use. However, the energy straits, incited by 1974's oil shock, showed that economies are dependent on energy. Turkey in particular was affected by this shock, as were other developing and newly industrialising countries. The period of recession continued until 1984, and the demand for energy from Turkey and other developing and newly industrialising countries increased around the late 1980s. This string of events contributed to the countries' need for new energy sources, and is still influential today.[229]

Turkey is dependent on outside sources on the issue of the security of its power supply, and almost half of its energy consumption is supplied by importing resources predicated on fossil fuel. The dependency on imported power brings along political weakness. In this context, Turkey's coherence with the EU's energy policy is important with regard to the diversity of power sources and enhancing the quality of these sources as well. Turkey gains leverage for both its strategic location and also its status a transit country for the transport of petroleum and natural gas. On the other hand, Turkey joins the Southeast European Power Market studies by its support of the EU's stabilisation program. Within this scope, the Regional Inter-Ministerial Meeting was signed in 2003 on the grounds of intercountry seizure. An international agreement has been signed by the EU. Acceding to a treaty for Turkey with the countries in the region, a supranational power community became constitutionalised; member countries are able to trade without being subjected to any restrictions in the subject of electricity, natural gas and the environment.[230]

---

229 EPDK (2011); Turkish Energy Market: An Investor's Guide.
230 MÜSIAD (2006); Türkiye'nin Enerji Ekonomisi ve Petrolün Gelecegi. Arastirma Yayınları, (Turkey's Energy Economy and Oil Future), Istanbul, p. 22.

Figure 17: Turkey Total Primary Energy Supply

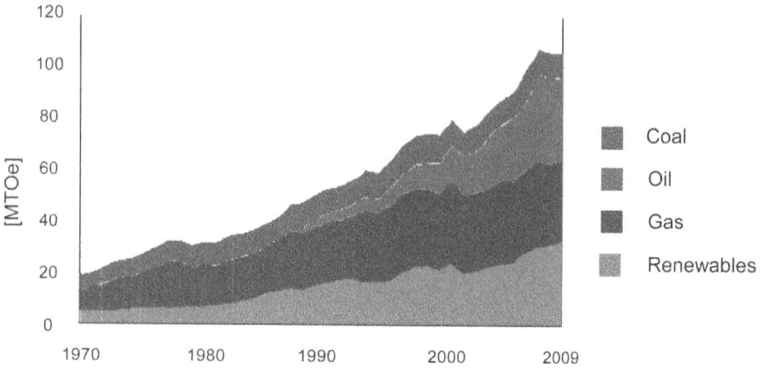

Source: ETKB, an Investor's Guide 2012

As can be seen in the figure above, Turkey obtained just 30.3 MTEP of its 106 MTEP primary power supply from national resources from 2009. As for the rest of its needs, they have been met through imported sources. Thus, Turkey is both an enormous energy consumer and an energy center, having a part in energy source distribution in its region. Since Turkey imports a big part of its energy, it conducts some activities to diversify the power sources and bring them to the country.[231] Turkey is a country that imports power from outside sources as it fails to meet its own power demands for energy generation capacity. Therefore, Turkey's energy generation has not risen to accommodate its rising population growth and, consequently, energy consumption growth. Therefore, the gap between energy generation and energy consumption has grown. The dependency on imported power sources is also higher in comparison to developed countries.[232]

BOTAS' (Turkish Petroleum Corporation) former general manager Fazil Senel said about Turkey's power supply security in his conference named *Turkey and Energy* that: "*We are working to minimize the foreign-source dependency for energy and to minimize the foreign-source dependency for natural gas and petroleum at the*

---

231 **EPDK (2011)**; Turkish Energy Market: An Investor's Guide, p. 7.
232 **Iktisadi Kalkınma Vakfı (2004)**; Avrupa Birligi'nin Enerji ve Ulastırma Politikaları ve Türkiye'nin Uyumu, (EU Energy Policy and Coherence of Turkey), Istanbul, pp. 53–56.

level 98%." Additionally, he said that: *"We are trying to reduce the foreign-source dependency for energy to around 50%."*[233]

Despite Turkey's predicament of dependency, its dependency rate of 57.8% on Russia in the uncertainty of energy security and available energy sources is in fact strategically compatible with the interests of the USA and the EU. Thus, these countries are collaborating on joint pipeline projects and infrastructure work. Turkey has to diversify its natural gas sources in case of any possible interruption related to the energy security of Russia to Turkey. Within this scope, Turkey has a tendency to diversify its natural gas sources with the neighbouring and power-rich countries such as the Caucasian countries, Central Asia and Caspian countries. The gas purchase agreements signed with Iran and Azerbaijan attest to this diversification. In this case, we see that the relationship between Azerbaijan and Turkey, according to the interdependence approach, *"consists of government and non-government organizations striving to increase their shares of profit from transactions, because they will reap benefits from this relationship, such as oil exporting governments and multinational oil companies. They share an interest in high prices for petroleum, however, they can be in conflict over the shares of the profits."*[234]

6.1 Rising natural gas use in Turkey

In the 1960s, Turkey was in the process of industrial constitution progress. As a result, its generation and consumption levels of energy were lower than they are presently. Previously, Turkey met a large proportion of its power requirements with renewable energy, and the share of renewable energy in the consumption of energy was over 40%. A large proportion of the power needs were being met by wood, animal and plant waste. The share of hydraulic sources in the generation of energy was 32% and petroleum was below 8%. However, primary energy generation increased at a rate of 4.3% and energy consumption increased at a rate of 6%; a 4% increase in energy consumption resulted from the intense industrialisation in the 1970s. Thus, the share of petroleum in total energy consumption has risen to 46.7% quickly. In the 1980s, the average increase of

---

233 http://www.haber3.com/botas-enerji-sektorunde-buyuk-oynayacak-745098h.htm (accessed on 30/01/2012).
234 **Keohane, O. Robert/S. Nye, Joseph (1989);** Power and Interdependence, Harper Collins, pp. 10–11.

energy generation remained at a rate of 2.2%; as for energy consumption, it rose at a rate of 4.4%.[235]

In particular, primary energy consumption, attributed to Turkey's growing and developing economy, has been on the rise recently. Turkey's energy consumption has been around 80.5 MTEP since 2009. Additionally, 194 billion kWh of electricity and 35.1 billion cubic meters of natural gas have been consumed in various sectors. As for the petroleum consumption, the crude oil level in 2009 was around 16.4 billion tons. Moreover, a quota per capita electricity consumption of 2865 kWh was the annual gross from 2009.[236]

The usage of gas has been constantly increasing when one examines the natural gas share in Turkey's energy imports since 1987. The share of natural gas in the total primary energy usage of 2010 reached the level of 32%, surpassing the petroleum level of around 6% in 1990. It is supposed that the natural gas usage level in Turkey will reach 82.7 billion cubic meters in 2020, as it was around 35.1 cubic meters in 2009. In the current conditions, natural gas is the most competitive source of residential heating in urban regions. As of 2011, 21% of the total imported natural gas, 8.7 billion m^3, was used in households for domestic heating. In rural areas, firewood is commonly used and approximately 6.5 million homes in Turkey utilise firewood as their primary heating fuel. As of 2008, firewood was the largest source of heat used for residential heating in rural regions. With that source of energy, biomass meets 5% of the primary energy demand in Turkey. The usage of solar energy is almost limited to water heating and amounted to 0.4 Mtoe in 2008.

According to the data, petroleum only met Turkey's energy requirement as far as 44% in 2010. Thus, Turkey imports 92% of its petroleum need and 97% of its natural gas. According to the data from the Ministry of Energy and Natural Resources, it is supposed that Turkey's total primary energy demand will reach the level of 222 MTEP by rising more than 100% in 2020, as the demand for electricity, natural gas and petroleum will reach the levels of 406-499 billion kwh, 59 billion cubic meters and 59 million tons.[237]

---

235 **Ege, Yavuz (2004);** Avrupa Birliğinin Enerji Politikası ve Türkiye'nin Uyumu"AB'nin Enerji Politikası Ve Türkiye, Uluslar Arası Politika Araştırmalar Vakfı, p. 29.
236 **EPDK (2011);** Turkish Energy Market: An Investor's Guide, p. 5.
237 Republic of Turkey Ministry of Energy and Natural Resources, Blue Book 2011.

Figure 18: Turkey's Natural Gas Imports

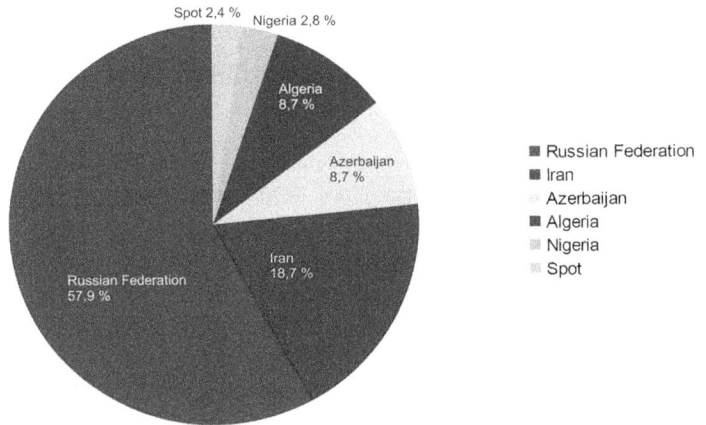

Source: Turkish Energy Market: An Investor's Guide, 2012.

As can be seen from the figure above, Turkey imports 57.9% of its natural gas demand from Russia, followed by Iran with 18.7%. In addition, Turkey has several LNG import agreements with Nigeria and Algeria. Imported natural gas is likely to be used in the production of electricity. When examining the sectoral allocation of national natural gas consumption in 2011, it can be seen that electricity received 48%, industry 26% and housing 25%.[238] Petroleum from the Middle East is imported from countries such as Iran and Saudi Arabia primarily, and then Russia. Finally, Turkey's demand for energy has risen at a rate of 8% annually. With this growth rate, Turkey has one of the biggest growth rates around the world, and has to generate solutions in the face of this rising energy demand.

Turkey would prefer Iraq's natural gas reserves to be developed. A clear benefit for Turkey is allowing Turkey to take advantage of the path of Iraq's natural gas through the Kirkuk-Ceyhan pipeline, as the pipeline that will be constructed in parallel with this pipeline will be connected to the Turkish national pipeline network. Within this scope, Turkey and Iraq signed a mutual agreement on 7 August 2007 in Ankara. It is estimated that Iraq's natural gas would reach Turkey and Europe via Turkey. The negotiation is still in progress based on the development of the present natural gas pipeline between Iran and Turkey. With the progress of these projects, Turkey aims to be the fourth main arterial road for Europe just after Norway, Russia and Algeria in the field of natural gas supply. Finally, it will

---

238 Turkish Energy Market: An Investor's Guide, 2012.

be monitored if the TANAP project's success, which has been signed between Azerbaijan and Turkey, will contribute to Turkey's position in this project.

6.2 Turkey's Dependency on Russia with Blue Stream Natural Gas Pipeline

The pipeline known as Blue Stream is a key subject for the power dialogue between Russia and Turkey. Turkey imports 57.9% of its natural gas requirement from Russia. A natural gas purchase agreement dated for 25 years was signed between Russia and Turkey on 15 December 1997, stating the transportation of Russian natural gas as 16 billion cubic meters per annum by the completion of the pipeline in 2002. It was planned that the amount of natural gas would be expanded to 51 billion cubic meters in 2013.[239]

Turkey ranks fourth among 19 countries importing natural gas through this pipeline and second in Europe, just after Germany. However, there are varying critiques concerning this pipeline. Turkey became two thirds prorate dependent on Russian energy at the start of this pipeline. Meanwhile, the transportation of Middle Eastern natural gas through Turkey has been blocked, which has paved the way for Ukraine to market its transportation. Additionally, it has become impossible to purchase the Turkmenian natural gas directly, which would be purchased cheaper than the Blue Stream natural gas. Besides, it was not approved to expand the pipeline to Lebanon, Syria and Israel, although two countries brought this issue to the agenda at the negotiations in 2000.

However, the negotiations still continue to expand the Blue Stream pipeline to the Mediterranean countries.[240]

6.3 Transmission of Gas Policy in Turkey

Turkey's energy demand is gradually increasing with its growing population and it only meets 30% of its energy requirement through internal resources. From 1960–1980, the state was dominant in the energy sector and the government was manipulating the marketplace at will. The structural reforms of liberalisation in the energy sector started in the 1980s, and the reforms in the economic field affected the energy policy on the same plane. As the liberalisation reforms in the energy sector rapidly increased, the state-controlled economy was replaced by a

---

239 Gazprom: Blue stream project between Turkey and Russia http://gazprom.com/production/projects/pipelines/blue-stream/ (accessed on 04/02/2012).
240 **Hacisalihaoğlu, Bilge (2008);** Turkey's natural gas policy; Energy Policy, June, 1870.

liberal economy, and the control of the government faded in electricity investments and the economy. Within this framework, the government has made legal regulations to draw the private sectors to the economy.[241]

The government constituted the energy market regulatory authority in 2001 to institutionalise the energy sector and to perform its supervisory mission. Especially, two statue laws introduced to the electricity and gas marketplace in 2001 have restructured these sectors, and there has been progress in these fields. Particularly, the electricity market law numbered 4626, enacted in 2001, has made many contributions to liberalise the energy markets and to create economic efficiency and environmental conservation. The electricity market law enacted in 2002 was another major contribution during the period of corporatisation. Twenty percent of the electricity market was opened to the private sector by this law, so that the competition and private investments were encouraged without any governmental guarantee. This law was changed in accordance with the European Union Electricity Directive in 2003. Furthermore, this regulation applied to the Electricity Sector Reform and Privatisation Strategy Document, which was discussed in 2004.[242] The obstacles of competition in the markets of electricity, natural gas, petroleum, and LPG and private sectors were removed thanks to the regulations from between 2001 and 2005. Thus, both domestic and foreign investors have been enabled to make their investments perspicuously and safely.

In addition to laws and regulations, the increasing demand and usage of natural gas in recent years have risen in importance in Turkish policy. The operations in natural gas transportation and trading in Turkey have been proceeding through BOTAS as a government enterprise. This institution, founded in 1974, has had crude oil and natural gas pipelines, and also a trading company. It was making the decisions on pricing policy before the natural gas law existed. However, the natural gas regulation law of 2001 abolished BOTAS's state of monopoly in the marketplace.[243] This law has allowed the energy market regulatory authority to start to take part as a competent authority in the competition for high-end natural gas pricing and supply. BOTAS, the monopoly corporation before the law, could not maintain its position after this regulation. After

---

241 Özer, Serra (2004); A Feasibility Study and Evaluation of Financing Models for Wind Energy Projects: A Case Study on Izmir Institute, A Dissertation, Izmir Teknoloji Üniversitesi, Izmir p. 16.
242 **Premiership Liberalization Administration (2004);** Electricity Energy Sector Reform and Liberalization Strategy Document.
243 **Foreign Economic Council (2006);** Energy Review.

the liberalisation, various energy companies such as Gaz de France, Shell, BP, Gazprom, Koc Group, Statoil, Eni, Pektim, Enerco and Avrasya Gaz Bosporus, Calık Group, Opet, Aygaz T Çont, Tupra, Petrol Ofisi and BOTAS started to carry out business in the gas sector.[244]

Within liberalisation studies of the power sector, a road map has been determined, starting within the perspective of full membership in the EU regarding the transitional period to the competition-based marketplace operation in the power sector. The Marketplace of Electrical Energy and the Supply Security Strategy Document have been accepted by the Higher Planning Council's decision dated 18.05.2009 and numbered 2009/11.[245] Within this scope, twelve distribution regions out of twenty-one have been conveyed to the private sector. Turkey is endeavouring to accord its power market to the EU by these changes and regulations. Regulating the power sector in accordance with EU standards, increasing cooperation and forming alternative power routes all contribute to the EU's power security in a positive way.

6.4 Turkey-EU Coherence after Negotiations in the Field of Gas Policy

On 17 December 2004, Turkey's accession negotiations began at the EU Council Summit. The EU commission mentioned the possible advantages of Turkey's probable membership results to the European Union. The Commission especially emphasised Turkey's role in the power issue. The importance of Turkey was also mentioned with regard to its geographical location and position as transit country in the route of petroleum and natural gas. Turkey's period of change began on a large scale at the start of Turkey's EU membership negotiations on 3 October 2005. Turkey's external dependence on energy is rising at an approximate rate of 72%, in parallel to its growing economy. Petroleum and natural gas in particular are completely imported from abroad and coal is partially imported. Thus, the energy cooperation founded between the EU and Turkey will increase Turkey's importance during the membership process.

It is estimated that this period will last at least ten years. Turkey has to complete 35 negotiation topics to adapt to EU standards within this negotiation period. Turkey has been endeavouring to complete the adaptation process in every sector within this period. The EU published a report in 2005 about Turkey

---

244 BOTAS International Limited: http://botasint.com/SirketProfili.aspx (accessed on 30/12/2012).
245 **ETKB (2011)**; Energy Ministry of Turkey, Blue Book p. 28.

as an EU negotiator country. The report stated that an international oil stock commission was founded within the scope of supply security in the section of power, and Turkey has to have an oil stock accounting for 90 days of the annual oil consumption specified by the Turkish International Power Agency. Furthermore, Turkey's adaptation of the security of supply was mentioned. So far, Turkey has had to uphold the law within the power affiance and its 2010 targets for renewable power sources during the negotiation period have not been reached.[246]

Despite these unreached targets, Turkey was given good feedback on the strengthening of supply security and power infrastructure investments in the 2008 progress report. Its supply security has improved, and the legislation to speed up the power infrastructure investments has been accepted. Apart from that, there have been some attempts to generate electricity by the private sector as well as the government according to the new law.[247]

There are some differences between EU member states and Turkey within both the policies and also the applications. The EU Commission makes some political decisions to make up for these differences by creating a common policy between the countries. For instance, the EU has postponed the marketisation of Poland's domestic market to foreign suppliers until Poland becomes a full member of the Union. It is known that Turkey's position as a transit country progressively affects the relations with the EU beyond the relations in the fields of oil and natural gas. Particularly, Turkey has strengthened its position by the latest developments regarding the Baku-Tbilisi-Ceyhan pipeline and TANAP. The Commission is more conscious of Turkey's important position, especially in recent years.[248]

The EU Commission has referred to developments concerning the topic of power in the 2010 Turkey Development Report. As another step of the project— the agreements with other countries—has been completed. Within this context, Turkey reached an agreement with Azerbaijan in 2010 in the matter of natural gas prices and its transit transportation through Turkey. Further cooperation between Azerbaijan and Turkey developed with the natural gas pipeline TANAP, the alternative to the Nabucco project. Moreover, Turkey has agreed with Italy and Russia on the construction of the Samsun-Ceyhan petroleum pipeline. Turkey has to have oil stocks in accordance with the EU and IEA directives and requirements.

---

246 European Commission, Turkey 2005 Progress Report, Brussels, p. 100.
247 European Commission, Turkey 2008 Progress Report, Brussels, p. 57.
248 Republic of Turkey Ministry of Foreign Affairs: Turkey Energy Strategy, May 2011.

There have been several developments to increase the stock, however, Turkey could not establish the required stock in the meantime.

By April 2010, Turkey had built natural gas lines to 66 cities, and two companies started LNG gas exports in 2009. According to the gas distribution programme, there had not been a development like this in this field.[249]

According to the IEA, Turkey does not have a national energy certificate in the energy field, but the Ministry of Energy has set a course for 2010-2014. Turkey is attempting to increase its domestic resources although its external dependence on petroleum and natural gas is increasing daily.[250]

In Turkey's harmonisation process to the EU, the liberalisation of the energy fields and power supply security are the major issues to mention. Particularly, the protection and privatisation of consumer rights in the liberalisation process should be clear. Liberalisation of the power market has been going on since 2001, as have the reforms in the electricity and natural gas fields. Within this scope, the Independent Energy Corporation (EMRA) and the licence regulation have been put into practice. Turkey has increased competition in the marketplace according to the electricity trade and power supply, and has begun to use the resources available to them.

The International Energy Agency (IEA) has indicated that Turkey is too slow regarding the privatisation of its natural gas sector, and the legalisation and regulation in the natural gas sector have been justified since the adaptation process in July 2008, although the natural gas market started in 2001. Thereafter, the liberalisation of the natural gas (LPG) sector began. In addition, BOTAS's monopoly still continued in the gas sector and there was no liberalisation of the import of wholesale pipelines.[251]

Consequently, the report by the IEA states that the reforms in the natural gas sector are too slow. This period needs to be accelerated within the power security. Besides, Turkey acceded to the IEA and has to count the capacity of natural gas and oil stocks over the long term. Thus, the advantages of using these sources in case of an emergency have been addressed. The EU Commission mentions that further cooperation between the EU and Turkey on the issue of energy would be mutually beneficial.

---

249 Turkey 2010-2011 Progress Report http://ec.europa.eu/enlargement/pdf/key_documents/2010/package/tr_rapport_2010_en.pdf (accessed on 06/01/2012).
250 IEA Energy Review Turkey 2009, p. 18.
251 IEA Energy Review Turkey 2009, p. 19.

6.5 Trans-European Energy Networks (TEN-E)

Another need for providing the security of energy supply is the reliable transportation of energy. In this context, the Trans-European Energy Networks (TEN-E) programme was initiated with the Maastricht Agreement for improvement in the fields of transportation, telecommunication, energy and the environment, in order to encourage the national networks to cooperate with each other. The aims were to contribute to the development of the domestic energy market, the improvement of energy supply security, and to provide benefit for the EU's economic and social adaptation. A foundation of sufficient energy connection networks was the goal, with the expansion of the domestic market by giving priority to the development of transfrontier natural gas and electric networks. As an example of this, the CENTREL (1995) electricity network, which includes Poland, the Czech Republic, Slovakia and Hungary, was connected to UCPTE, the primary electricity network in Europe. Research has been done in the context of the INOGATE (Interstate Oil and Gas Transportation to Europe) programme for former Soviet Union countries.[252]

6.6 Aims of Turkey's Gas Policy by 2023

The Prime Minister Recep Tayyip Erdogan explained that his party's 2023 election manifesto targets consist of five main topics. The Prime Minister outlined these topics as *"advanced democracy", "great economy", "strong society", "viable environment and brand cities"* and finally, *"leading country".*[253] Through these five areas, Turkey aims to be among the top ten economies in the world in 2023. Several goals have arisen amidst these targets, such as raising the external dependence in the field of energy, raising the potential of renewable power, and energy efficiency. Within this scope, the Ministry of Energy has made strategic plans for 2023. The Ministry of Energy and Natural Resources has explained the targets for 2023, stating that there will be at least five billion dollars of energy investment per annum, the installed capacity will be redoubled to 100,000 MW and the share of private sectors in the energy will be increased to 75%.[254]

---

252 European Commission: EUROPEAN PARLIAMENT AND OF THE COUNCIL of 6 September 2006 laying down guidelines for trans-European energy networks and repealing.
253 **ETKB (2011)**; Turkey Energy Policy, Ankara.
254 **Ibid.**

Turkey is located in an area that has 72% of the authenticated petroleum and natural gas reserves. According to the energy consumption projections for the future, it is estimated that the worlds' energy consumption will increase to 40% in 2030. In the East-West axis, the Baku-Tbilisi-Ceyhan crude oil pipeline became operative from the Ceyhan terminal in 2006 and shipped 800 million barrels of petroleum in total as of the end of 2009. The Ministry aims to double this amount in 2015 in comparison to 2008.[255]

The Ministry of Energy and Natural Resources wants to increase the procurement of electrical power from renewable resources in 2023, so it has planned to increase the procurement share of electrical power from renewable resources to the level of 30% in 2023. As for the current brute consumption amount per capita, it is approximately 2217 kWh in Turkey and 6602 kWh in EU countries. It is supposed that the Ministry will raise this amount to 5700 kWh per capita in 2020. It is required to increase the installed capacity of wind power from 802.8 MW in 2009 to 20,000 MW by 2023, and the currently installed geothermal power capacity from 77.2 to 600 MW by 2023.[256] More targets for 2023 are to increase energy efficiency, to prevent the wastage and to reduce energy density. To achieve these targets, the Ministry aims to reduce the primary energy density at a rate of 20% by 2023 in comparison to 2008. Within this scope, Turkey intends to provide a rapid and continuous improvement in its energy efficiency in order to reach an EU-standard level.

On the other hand, Turkey has aimed to provide resource diversification in the petroleum and natural gas sectors as its power import items. Thus, the consumption of natural gas was 36 billion cubic meters in 2008, as the natural gas generation was around one billion cubic meters that year. It is seen that the total rate of external dependency on natural gas is around 97.3%. In contrast, Turkey's petroleum generation in 2008 was 19.3 million barrels, and the external dependency in this field, in consideration of the consumption, was 93%. The Ministry of Energy and Natural Resources aspires to redouble the storage capacity of natural gas from 2009 to 2015 as the targets of these resources. As for natural gas, Turkey aims to reduce its maximum import share below 50% by 2015 at the latest. Concerning the environment, it aims to reduce the rate of greenhouse gas emissions caused by the energy sector after 2014. Finally, it is estimated that at least three nuclear power terminals will be completed by 2023.[257]

---

255   Energy and Natural Resources Ministerial 2010–2014 Strategic Planning p. 42.
256   **Ibid;** p. 22.
257   **Ibid;** p. 36.

6.7 Turkey's Energy Chapter in EU Membership

6.7.1. Renewable Energy Resources in Turkey

Turkey and the EU are more deeply dependent on Russia, the Middle East, and other exporting countries for gas supply than the United States, because Turkey has a larger vulnerability should the gas-producing countries ever limit the export of gas. Therefore, vulnerability is a source of political power in the interdependent world, which is why one of Turkey's aims is to reduce vulnerability to related parties and others. In other words, the sources of dependency must be diversified or decentralised to other sources, such as nuclear or renewable energy.

A law regarding the manufacture usage of renewable energy resources to procure electrical power was accepted on 29 December 2010. Enacted in 2001, this law has extended the direct price support for electrical power procurement from renewable resources until 2015. Further, there have been additional incentives for the usage of domestic equipment in the electrical power procurement from renewable resources.[258] According to the projections by the energy market regulatory authority, the current power potential of Turkey in renewable resources is around 136,600 MW. However, the amount Turkey is using from this potential is around 18,600 MW. The remaining amount of this potential (118,000 MW) is on standby mode. In the data of the energy market regulatory authority, it seems that Turkey's electricity requirement is approximately 212 billion kWh per annum, but it is estimated that this will increase to 500 billion kWh in 2023.[259] The analyst Arif Hepbasli explains in his work *Present Status and Potential of Renewable Energy Sources in Turkey* that: *"Turkey is a rich country in the sense of renewable power resources and it is a must to benefit from the solar power since there are many sunny days during the year. In this sense, the government promotion has to be arisen so that the required investments in this field could be made."*[260]

In short, current subsidies in the form of feed-in tariffs for electricity generation from renewable energy sources should also be introduced for heat production, in order to make the REHC technologies more competitive. Prior to that, analysis should be carried out with respect to different consumer segments, type and price of available biomass in the respective region, and the cost of the heating system being replaced. All these feasibility works should also take into consid-

---

258 **ETKB (2011);** Blue Book, p. 29.
259 ETKB, Documents about Nuclear Reactor Planning of Turkey, pp. 29–30.
260 **Arif, Hepbaşlı/Aydoğan Özdamar/Nesrin Özalp (2001);** Present Status and Potential of Renewable Energy Sources in Turkey, Energy Sources, Taylor & Francis, 23, pp. 631–648.

eration that weather conditions and heating needs vary widely between different regions in Turkey. Therefore, each region should be particularly investigated in terms of available renewable energy sources and energy needs. After undertaking the necessary investigations, support mechanisms in the form of feed-in tariffs, tax exemptions and reduced interest rates for loans, knowledge support and technological assistance should be developed and directed towards the promotion of renewable energy utilisation for heating and cooling.

6.7.2. Nuclear Energy Studies

The Ministry of Energy and Natural Resources claims that the rise of external energy dependence will increase in direct proportion to the population growth, and that it is a must to construct a nuclear energy power plant.[261]

According to the data submitted by the World Nuclear Association, the whole world has a total of 375.9 GW of power as of 2011, and there are 439 nuclear reactors in the world providing 13.5% of the world's energy generation. On the other hand, 61 nuclear reactors are under construction and 151 new nuclear reactors are planned to be constructed by 2022.[262] Within the context of this data, there are still ongoing nuclear energy power plants being constructed, even after the Fukushima nuclear power plant disaster. Turkey and Russia signed an agreement in Ankara on 12 May 2010 to construct and operate a nuclear power plant facility in the Akkuyu area in Turkey. Furthermore, there are ongoing negotiations with Japan to construct another nuclear power plant in Sinop. In an annotation prepared by the Ministry of Energy, it is stated that the security of energy supply will be provided, and the foreign-source dependency on energy and current account deficit will be reduced.[263]

France imports 99% of its petroleum demand and 97% of its natural gas demand from external sources, yet its external dependence on energy is around 50%, compared to Turkey's 72%. France meets this deficit by generating energy from nuclear power.[264]

With regard to nuclear power, MUSIAD, the businessmen's association, claim that the construction of a nuclear power plant is a must and that Turkey has to utilise nuclear energy to achieve its 2023 goal. It is estimated that Turkey's en-

---

261 **ETKB:** Documents about Nuclear Reactor Planning in Turkey, p. 2
262 **World Nuclear Energy:** http://www.world-nuclear.org/info/inf102.html (accessed on 05/0472013).
263 **ETKB;** Documents about Nuclear Reactor Planning in Turkey, p. 6.
264 **ETKB (2011);** Blue Book, p. 30.

ergy requirement will increase by 6.6% per annum during the next 20 years, and its expenditures on energy will increase accordingly. Turkey supplies 50% of its power generation from natural gas networks, but an additional investment valued at six to nine billion dollars annually is needed since the current account deficit and import incomes are used for energy consumption. Morever, the total energy procurement rates from nuclear power plants, when examining developed countries, are 78% in France, 54% in Belgium, 48% in Sweden, 37% in Switzerland, 32% in Germany and 30% in Japan.

Consequently, MUSIAD states that nuclear power should be considered, not only for energy demand and the security of power supply, but also in terms of its economic dimensions and environmental effects.[265] According to the security of power supply strategy paper of the Ministry of Energy and Natural Resources, it is aimed that the share of nuclear power plants in energy generation will reach at least 5% by 2020 and this percentage will rise in the long term.

6.7.3. Energy Efficiency Studies

Fossil fuels such as petroleum and coal are important energy resources that are consumed quickly. The greenhouse gas emissions emerging in the process of energy generation and consumption are one of the most significant contributors to global warming and climate change. There is 73% dependence on foreign sources to provide the security of energy supply. The risks based on this issue should be minimised; climate change must be minimised and avoided if possible, and energy efficiency rates should be increased. According to the data submitted by the Ministry of Energy and Natural Resources, there is distinguishable energy conservation potential in several fields, including 30% in the housing sector, 20% in the industrial sector, and 15% in the transportation sector. Energy efficiency legislation came into effect in 2007 with the objective of evaluating this energy efficiency potential and increasing this efficiency. The Ministry has set some targets so that the energy efficiency in the electrical power production plants and transmission and distribution networks will be increased and highly productive cogeneration applications will be extended. On the other hand, many rehabilitation works have begun in the Blue Book of the Ministry of Energy and Natural Resources within the context of energy efficiency to increase the efficiency and production capacity of the public thermal and hydraulic power plants.[266] Besides, it is required that the energy performance

---

265 http://enerjienstitusu.com/2011/06/15/musiad-nukleer-enerji-zaruri zorunluluk/ #ixzz1PNbNLjRq (accessed on 01/02/2012).
266 **ETKB (2011);** Blue Book, p. 32.

certificates prepared in the buildings with building permits adhere to a new energy efficiency law as of 1 January 2011. As for existing buildings, an extension period lasting until 2017 has been granted by which building permits must be obtained.

6.7.4 Climate Change Policy and Position of Turkey in the Kyoto Protocol

The Climate Change Framework Convention was signed at the Environment and Development Conference in Rio in 1992. According to this convention, different countries should shoulder different responsibilities by reckoning with their development levels. The developed countries have a responsibility to reduce emissions. Besides, they also have to support developing countries financially and technologically. As for developing countries, they are supposed to make an effort to reduce emissions. The energy sector, which causes more than 80% of the world's greenhouse gas emissions, is the key sector for the success of the policies and negotiations concerning climate change. Thus, the Kyoto Protocol was first opened in 1997. However, the countries that are responsible for more than 55% of the global greenhouse gas emissions must accede to the treaty for this protocol to be enured. The application of the protocol has been ineffective as the USA and Russia did not sign the protocol for a long time, however Russia accepted the treaty in 2005. This protocol has obligated the developed countries that acceded to the treaty to reduce emissions between 2008 and 2012 at the rate of 5% in proportion to 1990. As for the EU, it pledged that its member states would reduce greenhouse gas emissions at a rate of 8%.[267] The protocol has introduced legal regulations in order to limit greenhouse gas emissions and it also aims to provide an international emission tradership and technological and capital movements. As for Turkey, it acceded to a treaty for the United Nations framework convention on climate change in 2004. In 2007, it presented its national declaration. As a developing country, Turkey first hesitated and was reluctant to accede to the treaty and to sign the Kyoto Protocol. However, the Grand National Assembly of Turkey approved the commitment to the Kyoto Protocol on 5 February 2009. Thus, the Ministry of Energy and Natural Resources has set targets to secure a reduction in the rate of increase of greenhouse gas emissions caused by the energy sector after 2014.[268]

The Ministry has made some suggestions to minimise the loss caused by the procurement, transmission and distribution of electricity and to reduce the emis-

---

267 World Energy Council Turkish National Committee (2010); Energy Report, Ankara, p. 146.
268 **ETKB:** 2010–2014 Strategy Planning p. 49.

sions. Particularly, the rate of loss in electricity transmission is roughly 2.5–3%. However, the rate of loss in the electricity distribution is at a high level. Therefore, it is aimed to minimise the loss by privatising electricity distribution operations. The rate of loss in this field is around 15%. In OECD countries, this rate is around 4% to 6% across the globe.[269]

6.7.5 Challenges of Turkey's Energy Policy and EU Accession Process

The cooperation between Turkey and the European Economic Community was founded by the Ankara agreement signed on 12 September 1963, and the customs union agreement between Turkey and Europe has been in effect since 1996. However, Turkey was accepted as a candidate country in Helsinki in 1997. With this summit, Turkey gained the status of candidate country to be a full member of the European Union. Turkey has been trying to apply for the Copenhagen accession criteria in economic and political fields and it has also stepped up its application of its administrative capacity and legal coherence in many fields.

A partnership system covering twelve Mediterranean countries was founded for the first time in the field of energy between Turkey and the EU. The partnership between Europe and the Mediterranean was founded by the Barcelona Manifesto published in 1995, and has been referred to as the development of the power cooperation. Thus, Turkey has started to develop many projects together with the EU on a regional scale within this scope. Afterwards, it was decided to establish a European- Mediterranean form consisting of 27 common country representatives during the Trieste conference in 1997. In 1998, a European-Mediterranean power plan with the aim of developing the common partnership was created. Within this scope, the provision of coherence among the energy-generator and consumer countries in European and Mediterranean partners is set as a target. The Synergy Programme has been put into practice in the Mediterranean region within this scope. Twenty-six different projects within this programme that Turkey takes part in were enabled between 1994–1998.[270]

Turkey's commitments, brought about by being a candidate country in the energy field, were specified over the short and medium term in the accession

---

269 ETKB http://www.enerji.gov.tr/index.php?dil=tr&sf=webpages&b=enerji_cevre_iklim&bn=218&hn=&id=4303 (accessed on 05/04/2013).
270 **Cahit, Atlı (2002);** "AB Uyum Sürecinde Türkiye'de Enerji Entegrasyonu Çalışmaları" (Studies on Turkey's Energy Coherence on the Process of EU) Standard – Ekonomik ve Teknik Dergi, Ağustos, p. 73.

partnership document. These requirements were revised and accepted by the Commission on 14 April 2003 as follows:

- To form a programme so as to commit the regulations regarding the internal power market in particular.
- To supply convenient devices to make the authority in question do its duty and to enable the independence and effective operation of the supervisory authority for the electricity and gas sectors.
- To establish and provide an internal and competitive power market in line with the electricity and gas directives.
- To provide an advanced level of acquis alignment regarding the power efficiency and to empower the power economy applications.
- To start creating and applying a programme reducing the power dependency of the Turkish economy and boosting the renewable power source usage.
- To restructure the power services and the marketisation of the power market in accordance with the acquis; to strengthen administrative and regulating structure.[271]

The commitments asked over the medium term are as follows:

- To remove the limitations in cross-border trading.
- To complete the national legislation in accordance with the union acquis.
- To restructure the projects and power services in the TEN-Power directive principles in the European community and the marketisation of the power marketplace
- To strengthen the administrative and regulating structure.

The development of Turkey's adaptation to the European community's power acquis was referred to in the progress report published in 2003. In the report, it was stated that there have been many developments in the gas and power sector in terms of the competition capacity in the domestic power market, as there have been advancements in the fields of power efficiency and renewable power sources. Furthermore, it was also emphasised that the administrative

---

271 State Planning Organization, Secretariat General for EU Affairs, National Program of Turkey's Undertaking of the Acquis of the EU Ankara, 2003, p. 20 http://www.abgs.gov.tr/files/AB_Iliskileri/AdaylikSureci/IlerlemeRaporlari/Turkiye_Ilerleme_Rap_2003.pdf (accessed on 05/04/2013).

capacity of the energy market regulatory authority has become stronger day by day.[272]

Turkey has prepared a national programme that includes political and economic measures, designed to speed up the EU harmonisation process. Turkey has set some legal regulation targets, particularly; general rules of electricity and natural gas in the energy part of *"The National Programme for the EU's Acquis Commitment."* These targets include: adaptation to the internal power market, to retain the independence and efficiency of the regulating authority in the electricity and natural gas sectors; to remove the limitations in cross-border trading; to establish a competitive power market in accordance with the electricity and natural gas directives; to adapt the EU's power acquis out of the internal power market; to adapt the EU's acquis regarding the compulsory oil stocks; to prepare an acquis regarding power efficiency; to increase the power generation sourcing from the renewable power resources; and to encourage the projects classified as serving the common interests within the scope of the Trans-European Networks (TEN) Power directive principles to be applied in Turkey.[273]

Developing the EU cross-border natural gas and electricity networks has been prioritised. Within this scope, there have been several attempts by TEN to create only one power market in the EU, in accordance with the Maastricht Agreement. Thus, it has been assumed to develop EU TEN as covering the neighbouring countries. Turkey has attached priority to developing the projects within TEN. Turkey has some targets such as providing a connection between the Turkey-Greece Power Network interconnection and the Turkish Power System to provide a mutual and secure operation of the power systems by creating power connections between EU Networks and in third world countries, particularly the candidate countries within TEN. Robert Keohane, who focused his regime theory on cooperation in the world political economy, said that international coordination of policy is highly beneficial in an interdependent economy, but he maintains that cooperation in world politics is difficult. He argues this based on the need to analyse cooperation within the context of international organisations and shared beliefs. The patterns of cooperation through concepts and international regimes will enable actors to predict future patterns of interactions among states and plan proper economic arrangements and related political activities.

---

272 State Planning Organization, Secretariat General for EU Affairs, National Program of Turkey's Undertaking of the Acquis of the EU Progress Report of 2003 Ankara Ankara, 2003, p. 89 http://www.abgs.gov.tr/files/AB_Iliskileri/AdaylikSureci/IlerlemeRaporlari/Turkiye_Ilerleme_Rap_2003.pdf, p. 89 (accessed on 05/04/2013).
273 National Programme of Turkey for the Adoption of the EU Acquis, 2003, pp. 519–543.

Keohane emphasises that states should cooperate by communicating with each other and coordinating to adjust their policies so they can become significantly more compatible with one another. On the other hand, since the creation of international regimes involves cooperation, each state is expected to pursue its national interests and maximise its relative gains compared to others when negotiating an international regime agreement. An agreement has been reached on the principles of certain policies. The theory of complex interdependence is an alternative to explaining cooperation and state behaviour. According to regime theory, we see that Turkey is trying to adapt to EU legislative and executive prescriptions regarding energy policy. In this respect, it is predicted that Turkey will advance through changes of energy policy and also adjustments to the field of gas policy.

6.8 International Pipeline Projects as Energy Hub of Turkey

6.8.1. Baku–Tbilisi–Ceyhan (BTC)

The pipeline projects allow Turkey a critical opportunity to play a powerful role in both the regional and global energy market as a transit country. The Baku–Tbilisi–Ceyhan (BTC) Crude Oil Pipeline is one of the most important pipeline projects, since it carries around a million barrels of petroleum per day to the Mediterranean and is becoming an energy corridor. The length of the BTC pipeline is 1768 km and 440 km of it is within Azerbaijani borders; 260 km pass near Georgia-Tbilisi and the other 1076 km reach Turkey-Ceyhan. The 50 million tons of crude oil that will be annually generated in this region reach the sea terminal constructed in Ceyhan, and afterwards are delivered to the world market by tanker ships.[274] The pipeline constructed by the support of global and political powers and international finance corporations cost four billion USD. The BTC pipeline was inaugurated on 25 March 2006 and the first petroleum transportation reached Ceyhan on 29 May 2006. This pipeline continues to work through BP's cooperation with its share proportion of 30.1%. The petroleum coming from the BTC pipeline has also started being shipped to consumer regions.[275]

Analyst Necdet Pamir emphasises that: *"Turkey has been an energy corridor and an important transit country to meet the increasing energy demand of the EU*

---

274 Bakü-Tiblisi- Ceyhan Pipeline: http://www.bp.com/sectiongenericarticle.do?catego ryId=9006669&contentId=7015093 (accessed on 04/02/2012).
275 İşcan, İsmaik Hakkı (2007); Türkiye-AB İlişkilerinin Geleceği Açısından AB Enerji Güvenliği Sorunu, (Turkey-EU Relations in the Framework of EU Energy Security) p. 142.

by the construction of this pipeline. On the other hand, this pipeline has been really profitable for Turkey economically and for the security of the Straits. There also have been different political acquisitions for Turkey and it has strengthened Turkey's strategical and geopolitical situation. It came up with a new and alternative export line against Russia by the rise of source diversification.[276]

Figure 19: Map of BTC Pipeline

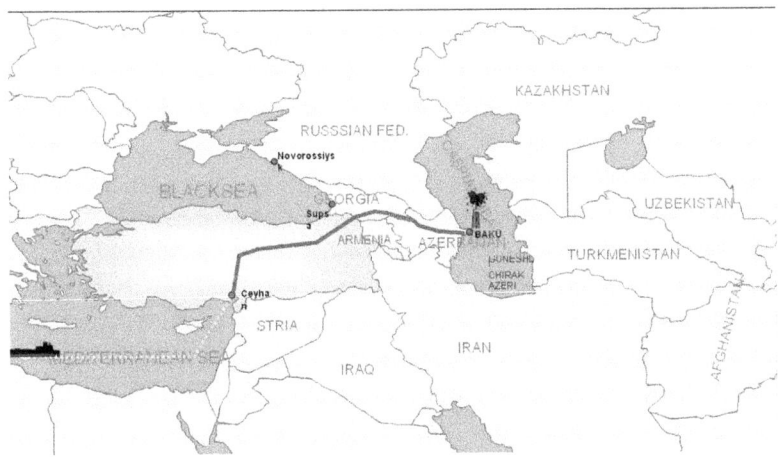

Source: BOTAS

As seen in the figure above, as a result of international projects, it is estimated that 6% or 7% of the world's petroleum demand will pass through Turkey, and Ceyhan will be an important distribution center. The East Mediteranean Sea will also play a significant role and be the biggest petroleum sale terminal. Turkey's strategic importance has increased by being an important passageway among the world's economic and power sources in the event that the crude oil coming from Kirkuk, Baku and Samsun finally reaches Ceyhan Terminal.[277]

Turkey endeavours to take a place in the projects that enable its energy sources to open up to the world market, a consequence of its developing relations with neighbouring countries in recent years. In particular, Turkey aims to make the

---

276 **Pamir, Necdet (2009);** AB'nin Enerji Sorunsalı ve Türkiye, (EU Energy Security Problems and Turkey) p. 81.
277 **Turkey Energy Strategy (2009);** Energy, Water and Environment, General Director of Turkey, pp. 5-6.

East Mediterranean the biggest power trading center by the pipeline projects that will connect to Ceyhan.

The first negotiations for the construction of the pipeline started in the 1990s; the project was approved and the World Bank Finance that supported the pipeline was given in 1998. The BTC was signed by the presidents of Azerbaijan, Georgia and Turkey, who came together in the organisation of a security and cooperation summit in Europe, where President Clinton was in attendance. It became official on 17-18 October 2000. This pipeline project has increased the geopolitical significance of Turkey. Besides, it is supposed that Turkey would generate an average income of around 200 million dollars during the first 16 years, and that this amount would increase to an average of 300 million dollars from the 17th and 40th years. A project of this magnitude can only be expected to positively affect the economy, and the construction of this pipeline gave employment opportunities to 20,000 people.[278] The construction of the BTC pipeline started in 2003 and was put into operation in 2006. In 2008, the transport capacity of the pipeline reached one million barrels per diem and its capacity increased to 1.2 million barrels per diem as a result of the capacity enlargement works to carry more petroleum through the pipeline in 2009, according to the data submitted by the Ministry of Energy and Natural resources. Turkey has been the key country of the East-West energy corridor through this pipeline. It is claimed that Russia has been oppressing several countries such as Azerbaijan, Turkmenistan and Kazakhstan to give up supporting the BTC and Trans Caspian pipeline projects, and is working against this project.[279]

While this project has provided economic benefits, on the other hand, there have been many arguments claiming that the BTC project is damaging the environment. It is also estimated that it will contribute to climate change and release one million tonnes of greenhouse gas emissions by the consumption of the carried sources.

6.8.2. The Samsun-Ceyhan Pipeline

The Samsun-Ceyhan pipeline is another important pipeline project of Turkey. It is planned that the petroleum of Russia and Kazakhstan would be transported first

---

278 **Küçükşahin, Ahmet (2006);** Türkiye'nin Enerji Stratejisi Ne Olmalıdır?, (What Should be Turkey's Energy Strategy?) Genel kurmay Başkanlığı Yayınları, pp. 128-129.
279 Republic of Turkey Ministry of Energy and Natural Resources: http://www.enerji.gov.tr/index.php?dil=en&sf=webpages&b=enerji_EN&bn=215&hn=&nm=40717&id=40717 (accessed on 05/04/2013).

to Samsun and then to Ceyhan via a pipeline set under the Black Sea. This project has been named Trans-Anatolia since it is a partnership of the Turkish company Calik Energy and the Italian company ENI. The above-mentioned companies signed a partnership agreement for the Samsun–Ceyhan Pipeline project in 2005. The groundbreaking ceremony for this pipeline occurred on 27 April 2007, and it was estimated that the pipeline would be 555 km long.[280] One million barrels of crude oil per diem and 50 million barrels of crude oil per annum were planned to be transported through this pipeline. Afterwards, two Russian companies, Rosneft and Transneft, got involved in this project through an agreement signed in Milan in 2009. Hakki Soylu refers to the importance of this project by mentioning it in his article: *"It is expected that the tank ship traffic in Turkish straits will be reduced considerably when this Project is completed. The petroleum of Russia and Kazakhstan will be opened to the world markets via Ceyhan so that Ceyhan will become a power center of Turkey. The city of Ceyhan will turn into an oil port just like Rotterdam by the completion of this pipeline so that the strategic importance of the country will be raised.*[281]

Finally, this pipeline would reduce oil traffic by bypassing the Bosporus Straits. Another project in planning, announced by Prime Minister Erdogan, is a 30-mile link between the Black Sea and the Sea of Marmara, tentatively called Canal Istanbul, and the waterway would be constructed by 2023 on the European side of the Bosporus. The aim of this project is to reduce oil traffic because of the threat of accidents on the Bosporus Straits.

6.8.3. Iraq–Turkey Crude Oil Pipeline (Kirkuk–Yumurtalik)

The Kirkuk–Yumurtalik pipeline brings the crude oil obtained from Kirkuk in Iraq and other production areas to the Ceyhan Sea Terminal. This pipeline was set into operation in 1976 and the crude oil began reaching the Ceyhan port in 1977. However, the pipeline was closed because of the Gulf Crisis in 1990 and reopened in 1996. Currently, only limited petroleum transportation is allowed, as per the decision of the United Nations' Security Council in 1997. The total amount of carried crude oil during 1997 was around 17–19 million tonnes. However, the political uncertainty and terror problems in the region brought along the pos-

---

280 Samsun-Ceylan Crude Oil Pipeline http://www.calikenerji.com/eng/rafineri.php?sf=rafineri_yatirim&k=283&u=422 (accessed on 04/02/2012).
281 **Soylu, Hakki (2007)**; "Rusya, Bulgaristan ve Yunanistan'ın Üzerinde Anlaştığı Burgaz-Dedeağaç Petrol Boru Hattı'nın Türkiye Açısından Önemi", Silahlı Kuvvetler Dergisi, Sayı. 391, p. 39.

sibility of an interruption in the pipeline. Between 1991 and 1995, there was no transportation of petroleum because of the politically motivated embargos placed on Turkey by Iraq. In 2010, it was estimated that the amount of crude oil carried via the Iraq-Turkey oil pipeline was 132,278 barrels.[282]

6.9 Important Export States

6.9.1 Iran

Russia is the second country after Iran in terms of gas reserves, and is the third biggest country for oil. The Nabucco project, which was planned, aimed at supplying gas from Iran. However, Iran is under international pressure and an economic embargo has been imposed by the USA on international trade with Iran. If the EU could have acted independently of the USA, Iranian gas could have been beneficial for the EU energy supply. In 1996, Turkey concluded a natural gas agreement with Iran for 25 years with a price of 20 billion USD. The pipeline was completed and the gas distribution started in 2001, despite the opposition of the US administration. Thus, natural gas is transferred from Tabriz, Iran to Erzurum through pipelines. In addition to this, Turkey annually supplies gas over pipelines of 10 bcm.[283]

There are reasons to be listed for selecting Iran as a route despite the US's opposition. In particular, Turkey first aimed at diversifying its energy supply and gas dependence. The fact that Iran is close to Turkey in geopolitical terms facilitated construction of the pipeline between the two countries and connection could easily be established with the countries of the region with rich energy reserves, both of which are considered as advantages.

On the other hand, Turkey signed an agreement in 2009 for extracting gas in Iranian gas fields, and thus Turkey has acquired the right to search and process natural gas in the Southern Persian basin of Iran. However, a gas cut was experienced during the winter period due to the political crises experienced with Iran from time to time. Therefore, periodic crises could arise in Turkey's gas supply from Iran.

---

282 BOTAS Petroleum Pipeline Corporation; http://www.botas.gov.tr/index.asp (accessed on 05/12/2012).
283 BOTAS Petroleum Pipeline Corporation; http://www.botas.gov.tr/index.asp (accessed on 05/07/2013).

6.9.2 Turkmenistan

Azerbaijan, Kazakhstan and Turkmenistan were the countries that produced gas in the Soviet Union. For that reason, Turkmenistan was known as the gas factory. However, Turkmenistan fell short of technological developments for oil and gas production. Thereafter, with the fall of the Soviet Union, there was political ambiguity during that period, which caused it to fail to obtain the investments required for domestic gas production and distribution. Meanwhile, it was dependent on Russia in many aspects, including its lack of financing and technological developments. For these reasons, the President of Turkmenistan, Turkmenbashi, led the PSA (Production Sharing Agreement) at no cost in order to increase the country's gas production potential.[284]

Turkmenistan has one of the five biggest reserves in the world. Due to the political and economic embargoes imposed by Western countries on Iran, it seems hard to provide gas from Iran to the Southern Gas Corridor. Therefore, EU countries are seeking ways to supply gas to the Southern Gas Corridor from Turkmenistan. However, the fact that the gas is to be produced in this country in the coming 20 years is dependent on Russia. Existing and long-term agreements make it hard for the EU and Turkey to export gas from the region. Some researchers are of the opinion that, given this framework, it is hard for the Turkmen gas to be supplied from the TANAP line.

Besides, the increasing energy demands of India and China demonstrate the interest in the region. Therefore, the interest in Caspian and Turkmen gas increases day by day. 1.2 billion m^3 of gas from Turkmenistan will be imported to China in 20 years. Accordingly, in 2009, gas import began in Central Asia and China. Pursuant to the agreement signed in 2008 between Gazprom, the Russian company, and the Turkmen government, an annual 50 billion cubic meters of gas will be connected to Russia. The EU wishes to be in close relations with the region. In 2007, in the Germany presidency, the EU Commission agreed on strategic cooperation with Central Asian countries. By developing relationships with these countries, they aimed to diversify and contribute to the EU energy supply and resources. The countries of this region host 4% of the global reserves and thus have a strategic importance for Germany and for the energy security of the EU. The German energy company, RWE, has signed a long-term memorandum agreement in the field of energy. Within this scope, RWE has opened an office in Ashgabat and started its activities to extract gas in the Caspian Sea.

---

284 **Götzö, Roland:** Die russisch zentralasiatische Energiegemeinschaft – eine Bedrohung für die europäische Energiesicherheit? S. 5.

On 20 October 1998, Niyazov, the president of Turkmenistan, and the then Prime Minister of Turkey Demirel agreed on the entry of Turkmen gas to Turkey and from there to the EU. This gas pipeline aimed to cover Turkey's gas need and then diversify the EU's gas supply. According to this agreement, Turkmenistan would supply 30 billion cubic meters of gas to Turkey, and 16 bcm thereof will remain in Turkey and the rest will be sent to the EU.[285]

The parties agreed to the Trans Caspian project in question for 30 years and the agreement was signed. The natural gas would be delivered from Turkmenistan to Georgia, and from there to Turkey.

On the contrary, the Caspian pipeline project was cancelled for three reasons. Initially, the transportation of the Caspian Sea project under the Caspian Sea was considered as high cost. The second is that the ambiguity of the legal status of the Caspian Sea was preserved. Finally, the Turkmenbashi Saparmurat Niyaziv's character was considered to be unreliable. Following the Blue Stream agreement signed between Russia and Turkey, the Trans Caspian agreement had no further importance. Following this, the energy ministers of the parties decided that it was difficult to cover both the Blue Stream project and the Trans Caspian project.

6.9.3 Iraq

Iraq has the third biggest oil reserves in the world. Iraq's gas and oil plays a significant role for Turkey in international politics. Particularly following the oil crisis of 1973, Turkey has changed its policies towards the Middle East. It was understood following the crisis that it had become harder to find cheap oil. Following this process, Turkey started to negotiate with Iraq and the Middle Eastern countries. The most well-known option at this point is the transfer of the raw oil from Kirkuk, Iraq to Ceyhan.[286]

The raw oil was first transported to Ceyhan in 1977 with tankers, and the parties agreed on expanding this pipeline between Turkey and Iraq.

The construction of the pipeline started in 1983 and was completed in 1984. Thus, the annual capacity of the pipeline reached 70.9 million tonnes or 1.6 mil-

---

285 Botas Petroleum Pipeline Corporation, Transcaspian Turkmenistan-Turkey-Europe Natural Gas Pipeline Project, http://www.botas.gov.tr/index.asp (accessed on 05/07/2013).
286 **Graham E. Fuller;** The New Turkish Republic (Washington DC: United States Institute of Peace Press, 2008), p. 39.

lion barrels.[287] In this regard, Turkey's biggest oil importer until 1991 was Iraq. However, the Iraq-Ceyhan pipeline was shut off for a period due to the United Nations embargo. It is estimated that the annual loss for Turkey for that reason was 1.2 billion USD.[288]

With the fall of Saddam in 2003, new negotiations were launched for the flow of raw oil to Ceyhan. However, it did not reach the level that it did before 1991. In 2009, 167,600 barrels of oil were transferred.[289]

Turkey's interest in the natural gas reserves in Iraq in particular has been continuing since 1990. Since 1996, Turkey has been negotiating on Iraq's natural gas reserves. The main purpose of the negotiation was to deliver Iraqi gas to Turkey, and then from there to other markets. In line with this purpose, Turkey aims to import ten billion cubic meters of raw oil annually through the pipeline.[290] Particularly following the fall of Saddam Hussein, the gas demand in the EU and Turkey increased. Thus, the negotiations were restarted on developing the natural gas reserves in Iraq. Consequently, Turkey and Iraq signed a memorandum agreement and agreed on extracting Iraqi gas and transferring it to Turkey and the EU.[291]

The oil and gas reserves of Iraq increase the role of Turkey in the international energy politics. It is estimated that the Iraqi gas reserves as evidenced in 2007 were 3.17 trillion. Seventy-two percent of these are located in the south of Iraq in the Shia region, and 14% are located in the north of the country. The remaining part is on the Syrian border. Turkish and US delegations carried out several negotiations on transferring the gas and oil resources in Northern Iraq to the EU. Until then, no success was accomplished in supplying gas from Northern Iraq due to various ethical conflicts between Northern Iraq and the central government.[292]

---

287 Botas Petroleum Pipeline Corporation, Iraq-Turkey Crude Oil Pipeline,http://www.botas.gov.tr/index.asp (accessed on 05/07/2013).
288 **Graham E. Fuller;** The New Turkish Republic, Washington DC: United States Institute of Peace Press, 2008, 39, 86.
289 Botas Petroleum Pipeline Corporation, Iraq-Turkey Crude Oil Pipeline, http://www.botas.gov.tr/index.asp.
290 **Ali Tekin and Paul A. Williams,** "Europe's External Energy Policy and Turkey's Accession Process", Center for European Studies Working Paper Series #170,2009, http://aei.pitt.edu/11786/01/CES_170.pdf. (accessed on 05/07/2013).
291 Republic of Turkey Ministry of Foreign Affairs, Turkey's Energy Strategy, January 2009, http://www.mfa.gov.tr/data/DISPOLITIKA/EnerjiPolitikasi/Turkey's%20Energy%20Strategy%20.
292 **Kramer Heinz (2010);** Die Türkei als Energiedrehscheibe, Wunschtraum und Wirklichkeit. S. 17.

In 2007, the Northern Iraq Kurdish Administration signed an agreement with seven big energy companies for gas production. Pursuant to the agreement, the energy company launched works for extracting and developing gas in Northern Iraq. In addition to this, the Austrian energy company OMW is engaged in works in Northern Iraq. Negotiations are ongoing for transporting Iraqi gas to Europe through the planned TANAP pipeline. Turkey intends to agree with the federal Kurdish administration and connect the gas resources of Northern Iraq to the TANAP project. At this point, Turkey also intends to transport the oil and gas reserves in the region to the EU. Developing its political contacts and trade connections will contribute much to Turkey, which is a safe country in the region. For Iraq, Turkey has the advantage of being both a consumer and a transit country. Iraq, on the other hand, has the opportunity of production charge and shorter transportation means. Iraqi gas is at a significant point for the Turkish gas supply. Meanwhile, economic cooperation with the EU as well as relations with Iraq will develop. In addition to this, it will provide economic contributions to Turkey in the form of transfer fees, and also contribute to EU energy security with regard to Turkey. The development of Iraq's natural gas could be considered as a strategic breakthrough for Turkey.

6.10 Southern Gas Corridor

The EU is becoming strongly dependent on Russian gas by the planned South and Nord Stream pipelines, which bypass Ukraine and Belarus and could be used by Russia to further its domestic and foreign interests. At the same time, the disputes between Russia and Georgia in 2008 were interrupted for three months because of the oil pipeline linking Baku with the Georgian port of Supsa, as a consequence Azerbaijani oil exports were damaged. Thereby, Europe began to pay more attention to the proposed Southern Gas Corridor to diversify their sources of natural gas imports. The South Stream will not threaten the realisation of the Southern Gas Corridor. According to Russian officials, Ukraine is an unreliable transit state for gas deliveries to EU states.[293]

Therefore, the security of the natural gas supply is at the top of the agenda for the EU. Most of all, Europe desires the security and diversification of supplies in order to decrease the level of vulnerability surrounding their dependence on Russian gas. The TANAP–TAP project, alternatively Nabucco, will deliver gas from the Caspian and Middle East regions to the EU market and provide the EU with

---

293 **Kucera, Joshua;** Armenian Military Simulates Attack on Azerbaijan's Oil http://www.eurasianet.org/node/66061 (accessed on 17/10/2012).

alternative export/transit routes. From Turkey, the Interconnector Turkey–Greece (ITGI) is the only pipeline transporting small volumes of Azerbaijani gas. Furthermore, according to the TANAP agreement, *"Ankara has the rights to transport additional gas to the Turkish market and will get a preferential rate for gas deliveries to Turkey. The pipelines connecting Turkey with Iran and Iraq oil and gas pipelines were damaged by rebels opposed to central government in Baghdad and by forces of the PKK. The BTC pipeline was closed for two weeks due to an explosion in Erzincan on August 2008. Despite Ankara's continuing problems over the Kurdish issues, Turkish politicians are negotiating with officials from the KRG to allow natural gas from northern Iraq to be transported to Europe via Turkey. Turkey plans to supply 15 bcm to the Turkish market with additional volumes to Europe."*[294]

The project was first mentioned at the Prague summit on 8 May 2009 as *"Southern Corridor – New Silk Road"*. The partner countries of the Southern Gas Corridor are Azerbaijan, Turkey, Georgia, Turkmenistan, Kazakhstan, Iraq, Egypt and Mashreg(Dubai), to whom the initiative was proposed by the EU. Furthermore, the EU has developed three of the pipelines that carry strategic importance such as ITGI, White Stream, and lastly TANAP-TAP. According to scientists, if the projects are realised, then it would be possible to deliver 60 to 120 bcm of Caspian and Central Asian gas to Europe per year.

Previously, the old version of the Nabucco project was planned to run from Azerbaijan to Austria. With TANAP, the pipeline has been revised to run from the Georgia–Turkey border to the Turkey-EU border. The TANAP project was initiated by Azerbaijan and Turkey and the EU is backing the Southern Corridor in order for Europe to receive Caspian gas, as well as Azerbaijani and also Turkmenistani gas. As mentioned, the TANAP route into Central Europe is shorter than the original Nabucco route. The Nabucco West and Trans Adriatic pipeline projects were completed, but the TAP was selected by Shah Deniz Consortium to divert the Caspian gas flow from TANAP. On this point, Azerbaijan's President Ilham Aliyec said, *"companies that have long been in Azerbaijan and are friendly to us can participate in this TANAP project as shareholders and investors. But, of course, we will determine which companies can participate in this project."*[295] In 2013, the consortium determined the plan for TAP with regard to Shah Deniz gas to Europe. Russia has developed and started the construction of the South Stream pipeline, which will bypass Ukraine and is an alternative to the Nabucco

---

294 Wing, Joel; "Iraq's Kurds' Gambit on Pipelines to Turkey may not Pan out" http://www.ekurd.net/mismas/articles/misc2012/5/invest841.htm (accessed on 29/05/2012).
295 **Socor, Vladimir**: Eurasia Daily Monitor Volume: 9 Issue: 148- August 3, 2012 Azerbaijan-Europe Gas Transportation Consortiums Face Major Restructuring.

project. According to researchers, this pipeline will increase dependence on that country. In any case, the EU must diversify energy supplies with pipelines such as TANAP in order to avoid dependence on Russia`s monopoly.[296]

The Southern Gas Corridor was also an important step in EU policy towards Turkey and potential gas suppliers. Turkey wants to become a long-term energy bridge between Ankara and the EU. The EU urges to initially secure an agreement on gas transit across Turkey. The Corridor could extend extra gas from Turkmenistan and, in the future, also from Kazakhstan as well as from the Middle East.[297] The difference between TANAP and Nabucco is that Nabucco had no shares in the oil company of Azerbaijan (SOCAR), which is why Nabucco revised its plan and created an alternative project. Firstly, Nabucco had financial difficulties in that time and SEEP had no other partners besides BP and a less-developed project. Additionally, the costs of building the pipeline were estimated at $12–14 billion. Secondly, Nabucco's capacity is 31 bcm, but ten bcm of gas are available.

The Shah Deniz I and II fields are not the only gas fields in the Azeri waters of the Caspian Sea. In the future, the Absheron field will be able to provide much more gas for Europe. Turkish and Iraqi supplies can significantly add to this supply diversity. Accordingly, Europe should obtain gas from shale sources from the continent in addition to gas from the Baku hub. Finally, the states are expected to take part in an interdependent relationship based on the potential benefits and costs.[298]

6.10.1 Azerbaijani-Turkish Project Trans-Anatolian Pipeline

There are differences between costly and beneficial interdependence. Minimising the costs of interdependence can lead to stability in the international system, allowing both parties to reap the maximum benefits of the relationship. Every interaction consists of benefits or costs of trade, and an interdependence analysis can point out these costs and benefits. Interdependence diffuses from interconnectedness because there is a mutual dependence between states. Thereby, interde-

---

296 http://www.jamestown.org/single/?no_cache=1&tx_ttnews[swords]=8fd5893941d69d0be3f378576261ae3e&tx_ttnews[any_of_the_words]=Nabucco&tx_ttnews[tt_news]=39403&tx_ttnews[backPid]=7&cHash=14f7efa1e66ef7e25b71f9df7b13dfff (Publication: Eurasia Daily Monitor Volume: 9 Issue: 98 (accessed on 23/05/2013).
297 Declaration–Prague Summit http://www.eu2009.cz/en/news-and-documents/press-releases/declaration---prague-summit-- southern-corridor--may-8--2009-21533/.
298 http://www.bilgesam.org/tr/index.php?option=com_content&view=article&id=2158:doalgaz-boru-hatlar-projelerinde-bueyuek-oyun-nabucco-gueney-akm-seep-ve-tanap&catid=131:enerji&Itemid=146 (accessed on 07/08/2012).

pendence emerges from interconnectedness, when there are transactions between states. Within this context, according to interdependence, Azebaijan would reach the gateway to the European gas market through Turkey, and on other hand, Turkey's energy hub position will be strengthened and will diversify through this project. Since both parties benefit significantly and need each other in order to do so, they share a mutual dependence. Azerbaijan and Turkey agreed at the memorandum on 11 December 2011 to establish a consortium, which proposes building the Trans-Anatolian Pipeline (TANAP) to supply gas from the Shah Deniz gas fields to Europe through Turkey. The President of Azerbaijan, Ilham Aliyev, and the Prime Minister of Turkey, Erdogan, signed an inter-governmental agreement on 26 June 2012 concerning the pipeline.[299]

However, despite this agreement, Turkey has suddenly decided to pass the South Stream pipeline through the Black Sea. Some researchers have criticised this decision. Yet the decision was not unfounded: the proposed Nabucco project uncovered some problems, mainly concerning finances and reserves. Therefore, Turkey was looking for alternative projects to diversify its energy resources. Azerbaijan and Turkey founded the TANAP project, which will change the situation and bring Caspian gas to Europe. On this subject, the EU Energy Commissioner Günther Oettinger mentioned with regard to the Trans-Anatolia Gas Pipeline (TANAP) that *"Europe is now a step closer to its aim to get gas directly from Azerbaijan and the other countries in the Caspian region. The EU supports the Southern Gas Corridor, which needs to build a new pipeline across Turkey. There are existing capacity constraints in Central Anatolia. In this regard, this pipeline will contribute both to Turkey's and the EU's security supply."*[300] TANAP is being supported by gas and funding from Azerbaijan. In this respect, the partners will finance the construction of the pipeline and their respective ownership shares, which are as follows: SOCAR 80%, BOTAS 10%, TPAO 10%. *"In the first stage, the transit line is projected at 16 billion cubic meters and will be increased to 24 bcm in the second stage. From this project, Turkey will be entitled to 6 bcm annually from Shah Deniz, in accordance with the long-term agreement between Turkey and Azerbaijan. This purpose project will be completed in 2017."*[301]

---

299 Direct Road to Europe: Azerbaijan's Trans-Anatolia Gas Pipeline Publication: Eurasia Daily Monitor Volume: 9 Issue: 2 January 4 2012, http://www.jamestown.org/single/?no_cache=1&tx_ttnews[tt_news]=38834 (accessed on 05/04/2012).
300 http://europa.eu/rapid/press-release_IP-12-721_en.htm?locale=en (accessed on 10/09/2012).
301 http://www.euractiv.de/energie-und-klimaschutz/artikel/gaspipeline-deal-tuerkei-und-aserbaidschan-fr-tanap-006466 (accessed on 02/03/2013).

Furthermore, Turkey and Azerbaijan urge maximising TANAP's gas flow to reduce dependency on Russian gas supplies. Hence, both countries want to supply extra resources other than Azeri gas. In this sense, Turkmenistan's Energy Minister Myrat Artykow, Azerbaijan's Industry and Energy Minister Natiq Aliyev, the European Union's Energy Commissioner for Energy Günther Oettinger, and Yildiz negotiated the TANAP project and how it would include Turkmen gas. If it included Turkmen gas, then TANAP's annual capacity could reach 60 bcm, initially starting at 16 bcm of gas. In this way, Turkmenistan is looking to ease its export dependency to Russia and expand. Turkish Energy Minister Taner Yildiz in this case said, "*We are a candidate to buy Turkmen gas.*"[302]

Figure 20: Map of the TANAP pipeline

Source: TANAP consortium

The control of the Shah Deniz II gas exports via the TANAP pipeline will lie with Azerbaijan itself rather than with a European pipeline consortium. At first, TANAP will be able to directly sell its own gas to European customers through its pipeline via Turkey's western border with the EU. As it is an Azerbaijani-owned pipeline it will not have to pay for the transit service, making Azerbaijan's gas price competitive in Europe. Aliyev defines this project as the "*direct road from Azerbaijan to Europe*".[303]

---

302 http://eurodialogue.org/Big-Game-heating-up-on-Caspian-gas-with-Turkey-Turkmen-move (accessed on 03/04/2013).
303 Direct Road to Europe: Azerbaijan's Trans-Anatolia Gas Pipeline Publication: Eurasia Daily Monitor Volume: 9 Issue: 2 January 4 2012, http://www.jamestown.org/single/?no_cache=1&tx_ttnews[tt_news]=38834 (accessed on 03/04/2012).

Turkey's role as an energy corridor to Europe and the intersection of multiple supply routes in Turkey take an important place in the TANAP project. Furthermore, TANAP stores gas in Turkey, making Turkey even more integral to the project. Moreover, the Turkish EU Minister stressed that *"TANAP will not hinder construction of the Nabucco natural gas pipeline"*, stating that *"[t]hose who want to take natural gas through Turkey will be able to build their own pipelines starting from the Turkish-Bulgarian border to meet their demands."* According to

Turkish Energy Minister Taner Yildiz: *"TANAP has become a project that regenerates Nabucco, instead of destroying it. Nabucco will be Nabucco if it's completed. But if the 2050 mile-long (Nabucco) line cannot be completed, then maybe we should prioritize the 800 mile-long (TANAP) line. TANAP will subsequently be the main artery of the pipeline network."*[304]

Evidently, TANAP will become an integral basis for the Southern Gas Corridor and take over the functions of Nabucco's Turkish section, though alternative routes will need to be considered. The pipeline is estimated to be the best option to connect Turkey to European markets.

6.10.2 Competing Existing Pipelines around Turkey's Border

6.10.2.1 Short Version of Nabucco

An agreement was signed for the construction of this pipeline within the Nabucco Project on 11 October 2002. The base for this project was first laid in 2002 as a result of the negotiations of BOTAS with Austrian, Bulgarian and Romanian gas companies. In 2003, the EU TEN Finance Committee agreed to support the project by granting 50% of the project feasibility cost. The European Council signed a joint venture agreement to put the project into practice by a resolution on 9 March 2007. With this agreement, an international company named "Nabucco Company Study Pipeline GmbH" was established by five companies, with headquarters in Vienna.[305]

Nabucco was an important project aiming to transport the natural gas reserves from the Caspian region to European markets through Turkey. Besides, Nabucco had the feature of being the biggest natural gas pipeline project among

---

304 http://www.upi.com/Business_News/Energy-Resources/2012/07/05/TANAP-pipeline-to-bring-gas-to-Bulgarian- border/UPI-83161341502061/ (accessed on 05/07/2012).
305 NabuccoGasPipelineCompany:http://portal.nabuccopipeline.com/portal/page/portal/en/company_main/about_us (accessed on 31/01/2012).

the Middle East, the Caspian region and the EU. This pipeline stretched from the eastern borders of Turkey to Baumgartner in Austria, making it one of the most important pipelines in Eastern Europe, Bulgaria, Romania and Hungary. It was estimated that the 3900-km-long pipeline would transport 31 cm natural gas by the completion of the project.

An agreement signed in Ankara in July 2009 supported the construction of the pipeline.[306]

Figure 21: Map of Nabucco West Gas pipeline

Source: Nabucco Gas Pipeline Company

The Nabucco gas pipeline, desired by the EU, is likely to become the shortened route to Turkey and aims to cover over 1300 km. Azerbaijani President Ilham Aliyev and Turkish Prime Minister Tayyip Erdogan decided on the shortened version, the construction of which cost 5.6 billion EUR. The aim of the pipeline was to transport 16 billion cubic meters of gas through Turkey, which could provide ten bcm of gas to Europe. The operator of Azerbaijan's Shah Deniz II announced that they wished to use the shorter liner as a preferred route for sup-

---

306 Nabucco Gas Pipeline Company: http://www.nabuccopipeline.com/portal/page/portal/en/Home/the_project (accessed on 31/01/2012).

plies to the Nabucco West, however, Shah Deniz gas producers' consortium in Azerbaijan announced its selection decision in the end of June 2011. Accordingly, the TAP project selected by consortium owes their existence to the Azerbaijani-led Trans-Anatolia Pipeline Project (TANAP), which is planned to run from the Georgian–Turkish border to the Turkish–European Union borders with Greece. After this decision, Gerhard Roiss, CEO of Austrian state oil company (OMW) stated, regarding the future Nabucco project, that "*The Nabucco project is over for us. Our goal now is European gas for European customers*".[307]

Nabucco was a long-term priority for the the EU Commission. However, their focus is now on the Nabucco-West and the TAP projects. The Commission views the opening of the Southern Gas Corridor as an important alternative.

6.10.2.2 Trans Adriatic Pipeline

The Trans Adriatic Pipeline (TAP) is planned to transport ten bcm per annum between Italy, Albania and Greece; all three countries agreed with the construction of the pipeline. TAP will start in Komotini/Greece and has been selected by the consortium for gas transportation in 2012. The final phase was decided on 28 June 2013, the route being from Greece via Albania to Italy. Commission President José Manuel Barroso and Commissioner Oettinger attended in Baku and, with President Ilham Aliyev, agreed that Azerbaijan will be "'*the substantial contributor' and 'enabler' to Europe's future gas deliveries from the Caspian region.*"[308]

Figure 22: Map of the TAP Pipeline Route

Source: Trans Adriatic Pipeline

---

307   http://www.eurodialogue.org/taxonomy/term/70 (accessed on 01/09/2013).
308   http://europa.eu/rapid/press-release_IP-12-1041_en.htm (accessed on 02/03/2013).

TAP would bring Azerbaijani gas primarily to the Italian gas market. At the same time, it proposes delivering some of that gas to Germany using Italian pipelines to the Swiss transit pipelines, Transitgas, which would connect with the German grid. Moreover, if TANAP selects the route, TAP intends to deliver gas volumes to Albania. In this case, a memorandum of understanding was signed in 2011 for small-volume deliveries to be sent to Bosnia and Croatia. TAP claimed a chance to diversify supplies in the Balkans or Southeastern Europe. The TAP consortium includes Norway's Statoil and Swiss Elektrizitaetsgesellschaft Laufenburg (EGL) with stakes of 42.5% each, and German E.On Ruhrgas with 15%. TAP offers the hypothetical possibility of a storage site in Albania.[309]

In this regard, the EU Energy Commissioner Günther H. Oettinger stated: *"We have a definite commitment from Azerbaijan that gas will be directly delivered to Europe through a new dedicated gas pipeline system. Whether the system consists of two gas pipelines – TANAP and TAP – or one single pipeline as earlier projects had foreseen – does not make any difference in terms of energy security. We now have a new partner for gas, and I am confident that we will receive more gas in the future."*[310]

As mentioned in previous parts of this dissertation, the EU is negatively affected by the crisis between Russia and Ukraine, and it seeks to diversify its energy sources and not depend on only one country for its energy needs. This project, then, would be beneficial to the EU. This would guarantee energy supply and the transport of natural gas from the Caspian region and the Middle East to Europe without the need for cooperation with Russia or Ukraine. The TAP Project is not only a power cooperation between the EU and Turkey, but it has also contributed to EU–Turkey negotiations as an indication of strategic partnership.[311] This project contains fundamental importance, as it can diversify the power supply over the long term and improve the EU's energy security. Several countries, such as Iraq and Egypt, demand to participate in this project.

---

309 http://www.europeanenergyreview.eu/site/pagina.php?email=faik_61%40hotmail.com&id_mailing=302&toegang=577bc c914f9e55d5e4e4f82f9f00e7d4&id=3455 (accessed on 12/10/2012).
310 http://europa.eu/rapid/press-release_IP-13-623_en.htm?locale=en (accessed on 05/08/2013).
311 **Laciner, Sedat/Özcan; Mehmet /Bal, Ihsan (2005);** European Union With Turkey: The Possible Impact Of Turkey's Membership On The European Union, Ankara: USAK Books.

Because of ongoing political crises, however, Western countries are not accepting of Iran.[312]

6.10.2.3 South East Europe Pipeline

The proposal for the South East Europe Pipeline was firstly for a natural gas pipeline from eastern Turkey to Austria, had the original Nabucco project been built. It was intended to diversify the natural gas delivery routes for Europe from Azerbaijan and to supply Europe with ten bcm natural gas. However, the Nabucco project was changed and revised as TANAP. In this sense, the consortium has selected TAP as an existing pipeline route from western Turkey to Europe. So far, it has preselected the Nabucco West pipeline. The South East Europe Pipeline would carry Azeri gas through Turkey to European markets, where there are competitors in the Southern Corridor such as Nabucco West, TAP and ITGI. All these projects pose challenges in varying degrees to pre-existing pipeline projects. According to the decision of the consortium, the offer of the competitor South East Europe Pipeline, owned by BP, has been turned down.

The South East Europe Pipeline (SEEP) was seeking the changes to be an existing pipeline aimed at cutting investments costs and adjusting pipeline capacity to the gas volumes guaranteed in Azerbaijan. It argued that usage of existing gas infrastructure in Europe was the more viable economic and political option for the Shah Deniz Consortium.[313]

6.10.3 Baku–Tbilisi–Erzurum

The Baku–Tbilisi–Erzurum natural gas pipeline is another project transporting the energy resources from the Caspian region to Turkey. It aims to transport 8.8 bcm of natural gas through this pipeline, in parallel to the BTC crude oil pipeline. The Baku–Tbilisi–Erzurum pipeline began to transport the natural

---

312 **O'Rourke, Breffni (2006);** Caspian: EU Invests In New Pipeline. Radio Free Europe / Radio Liberty http://www.rferl.org/content/article/1069500.html (accessed on 15/03/2013).
313 Jamestown: South-East Europe Pipeline: A Downsized Nabucco Proposed by BP Publication, Eurasia Daily Monitor, Volume 8, Issue 202. http://www.jamestown.org/single/?no_cache=1&tx_ttnews[tt_news]=38609&tx_ttnews[backPid]=228&cHash=ce4cc1f73f4e7c8be75725182a346259 (accessed on 09/10/2012).

gas in November 2006, and it aims to expand this pipeline to Kazakhstan and Turkey.[314]

On another issue, Iran is the second largest natural gas importer for Turkey, just after Russia. The construction of the pipeline started after the agreement between Tabriz and Erzurum was signed in 1996. The pipeline has been put into operation by transporting 20 bcm of natural gas per annum just after the completion of the pipeline construction in 2001. It is aimed to connect this pipeline to the Baku–Tbilisi–Erzurum pipeline.[315]

6.10.4 South Europe Gas Ring Interconnector Turkey–Greece–Italy Pipeline

The Caspian Sea is an important project that would transport the natural gas coming from the Middle East and South Mediterranean countries to EU countries through Turkey. This project is supported by the EU, which involves Turkey, Greece and Italy. The groundbreaking ceremony of this natural gas pipeline was done by the Primer Ministers of Greece and Turkey in 2005, and the pipeline between these two countries was put into operation in 2007. The pipeline between Greece and Italy is to be expanded and eventually reach Europe.[316] The ITGI project is supposed to supply Azeri gas to southern Europe through the Greece–Bulgaria Interconnector.

6.10.5 Azerbaijani-Turkish Natural Gas Pipeline Project (Shah Sea I and II)

The Shah Sea I project was signed as an agreement for the natural gas purchase between Turkey and Azerbaijan on 12 March 2001. Its objective is to raise the transportation of natural gas to 6.6 billion cubic meters per annum from three billion cubic meters. Turkey has begun to receive Caspian natural gas with a 15-year dated agreement made for this pipeline. The Turkish part of the pipeline is planned to be 224 km long. The first natural gas was generated by the Shah Sea Project in

---

314 Republic of Turkey Ministry of Energy and Natural Resources: Natural Gas http://www.enerji.gov.tr/index.php?dil=tr&sf=webpages&b=dogalgaz&bn=221&hn=&nm=384&id=40694, (accessed on 14/11/2011).
315 **Roberts, John (2004)**; The Turkish Gate Energy Transit and Security Issues, CEPS EU-Turkey Working Papers, No. 11, p. 53.
316 Republic of Turkey Ministry of Energy and Natural Resources: Natural Gas http://www.enerji.gov.tr/index.php?dil=tr&sf=webpages&b=dogalgaz&bn=221&hn=&nm=384&id=40694 (accessed on 01/10/2011).

2006. This pipeline has also made important contributions to the BTC Project, and would reduce the costs since it has been set parallel with the BTC pipeline.[317]

Another important agreement between Turkey and Azerbaijan is the Shah Sea II project. This agreement was signed between Turkish Prime Minister Recep Tayyip Erdogan and Azerbaijani President Ilhan Aliyev in 2010. According to this agreement, the transportation will start at two bcm in 2018 and will increase fractionally to four cm in 2018 and six bcm in 2019. It is believed that the Shah Sea I and Shah Sea II projects between Azerbaijan and Turkey have increased the chance for the TANAP Project to be brought into action sooner. Thus, Turkey will be an important resource in 15 years with regard to natural gas supply and, in effect, supply security. Concerning the project, Aliyev stated that "*approximately 16 bcm natural gas in Shah Sea II Project would be added to the 9 bcm natural gas in Shah Sea I so that Turkey will keep the aforementioned 16 bcm natural gas it needs within the scope of the security of energy supply.*"

Besides, it is accepted by the parties that Turkey will have the opportunity to commercialise 1.2 bcm of the natural gas to third party countries. By this agreement, the Shah Sea I project has been regenerated since 2008 and the cost, amount and transit fares of the Shah Sea II project that will become a part of the activity in 2016–2017 have been regularized.[318]

---

317 **Anar Hüseynov (2002);** Gas Perspectives of Azerbaijan, Economic News Bulletin, No. 3.
318 Shah Deniz Project :http://www.haber7.com/haber/20100607/Azerbaycanla-Sahdeniz-projesi-imzalandi.php (accessed on 07/06/2010).

7. Energy as a Factor in the Relationship between Turkey and the EU

7.1 Turkey's Geostrategical Factor

Today, rational and pragmatic policies consist of actors in world politics in interdependent relationships. The globalisation effect is one reason why it is required to enter transnational and international transactions. In the last years, Turkey's strategy and foreign policy is becoming concerning, and the country is considering a number of alternatives with their power sources.

As it was indicated in Chapter 3 of this research, Turkey's strategic importance has increased due to its status as a neighbour to the countries that own 70% of the global natural resources as well as to an energy corridor for the EU. Naturally, Turkey with its political stability and growing economy in recent years, is one of the most reliable countries in the region. The projects it has developed have increased this geopolitical importance. Finally, the TANAP project, which has been launched with Azerbaijan as an alternative to the Nabucco project, is a good indicator of this. With the addition of Turkmenistan and Iraqi gas to this project, a significant step will be taken in reducing dependence on Russian gas. As it was demonstrated in the study, the fruition of this project has the potential to increase interdependency between states and thus ensure economic and political stability for Turkey.

Turkey gained a huge geostrategic importance in international economic power just after the Middle East countries struck oil, and although it has been considered as infertile and unimportant soil, when considering Turkey's relations with countries in the Middle East, it has started to play a crucial role in the geopolitical and geostrategic junction area—a significant conflict area in the modern era. In particular, the Middle East has gained strategic importance through Europe's rising demand for power resources after the Industrial Revolution, since its position for trade and resource transmission has turned it into an out-and-out natural resource reserve. Both the region's strategic position and its containment of an important part of the petroleum reserves have caused intense competition between the regional powers.[319] Its importance is especially attributed to the fact that five of the most important sea passageways in the world pass through it (Bosporus and Dardanelles, the Suez

---

319 **Davutoglu, Ahmet (2001);** Strateji Derinlik, Türkiye, nin Uluslararasi Konumu, (Strategic Depth: Turkey's International Position) Küre yayinlari, p. 333.

Canal and the Strait of Hormuz and Aden).[320] The support of some Western powers to Israel in the Middle East has instigated a big reaction from other Middle Eastern countries. Notably the Saudi Arabian King Faysal and the other regimes have gravitated to a common reaction, in which petroleum is now used as a regional strategic trump card. Davutoglu states in this regard that the: *"Organization of the Petroleum Exporting Countries (OPEC) has gained an international power by the oil embargo during 1974. It has seen the influence of the Western countries' dependence on the natural resources by the oil embargo in that period of time so that Turkey has started to adopt more directive and determinant policy regarding the Middle East policies. On the other hand, EU countries are less dependent on the petroleum in comparison with the USA. However, the USA is aware that it would not be in the international economy and policy competition unless it has a control over the countries having the petroleum reserves.*[321] Davutoglu mentions in his work *Strategical Profoundness* that two major events developed at the end of the 1970s that triggered a strategic oil-based competition. First of all, the Camp David agreement signed under the leadership of the USA broke efficiency in the Middle East. The war between two major oil-producing countries, Iran and Iraq, caused OPEC to use petroleum as a strategic trump card on the other side.[322]

In the light of the developments occurring in the Middle East, Turkey did not develop any policy for the region although it was not playing an effective role in the oil-based balance. On the contrary, it suffered a severe blow during the oil shock in the 1970s and was one of the most economically endangered countries in the Gulf War during the 1990s.[323] Consequently, the importance of the Caspian Basin and the Middle East Region has increased since the North Sea, the only oil field the Westerners have, will be consumed in the near future. Within the scope of geopolitical strategy, the exterritorial powers and multinational companies became involved in the pipeline competition between Turkey and Russia. Thus, the Baku-Ceyhan, BTE and TANAP-TAP projects have been the most important step of the East-West energy corridor, a corridor planned to be connected to the Caucasus Southern, Central Asia, Turkey and the Mediterranean Sea.[324] Currently,

---

320 **Ibid;** p. 327.
321 **Davutoglu, Ahmet (2001);** Strateji Derinlik, Türkiye, nin Uluslararasi Konumu, (Strategic Depth: Turkey's International Position), Küre yayinlari p. 334.
322 **Ibid;** p. 334.
323 **Ibid;** p. 335.
324 **Baskin ORAN (2010);** Türk Dis Politikasi Cilt II: 1980-2001 Iletisim Yayinlari (Turkish Foreign Policy), (E.d) **Mustafa Aydin** (Kafkasya ve Orta Asyayla Iliskiler) p. 431-437.

there are economic and military interests due to globalisation. In this sense, the actors try to increase collaboration. However, a major threat is posed by actors who are largely outside of socially accepted interdependent relations because they operate independent of normal political, economic and military structures.

7.2 Rising Importance of Turkey in European Energy Security

After China, Turkey is the second most demanding country of natural gas and electricity in the world in the last ten years. This boom is parallel to the economic and social developments. Turkey aims to be the fourth main arterial road in the natural gas supply of Europe just after Norway, Russia and Algeria. Thus, they try to increase a new cooperation by gaining strong Turkey–EU relations. As for the work on the Blue Stream natural gas pipeline in the North–South axis, the Samsun–Ceyhan oil pipeline and Turkey–Israel Energy Corridor, these projects are still going on. With the completion of these planned projects, it is estimated that 6% or 7% of oil would pass through Turkey, and Ceyhan would be an important power distribution center and the biggest oil sale terminal of the East Mediterrian Sea. Turkey is one of the most important and irreplaceable partners to meet the increasing power demand of the EU and to diversify the resources by both its geopolitical location and strong infrastructure in comparison to the neighbouring countries and their rich natural resources.

The power policies have to consider the nations' national and economical security as they set the governments' objectives of foreign policy. The power security underpinning the power policy in particular was affected by the oil shock during the 1970s and power security was devastated by the dissolution of the Soviet Union, the Gulf Wars and then the September 11 attacks. After the Cold War in particular, Turkey's position in the international field was strengthened by Eurasia's and Turkey's gaining importance within its strategical and geopolitical location.[325] On other hand, Turkey is a reliable state and is a member of several western institutions such as NATO, OECD and Customs Unification with the EU. The other advantage of Turkey is that has large capacity refineries, underground natural gas storage and oil pipeline networks and the Ceyhan terminals as its international energy terminal.

A large part of the EU's power sources, the most important being the oil fields, are located in the Arctic Ocean. However, there is another reason for

---

325 İscan, İsmail Hakkı (2007); Türkiye-AB İlişkilerinin Geleceği Açısından AB Enerji Güvenliği Sorunu (Turkey-EU Relations in the Framework of EU Energy Security), p. 136.

the Caspian Region to gain importance day by day; there will be a depletion of reserves in the near future. Thus, the Caspian region has a big importance due to its power resources. The power sources in this region need to be carried throughout the world by new transportation routes. Therefore, Turkey and Russia are two competitor countries that want to take charge of the bridge. Russia uses the power policy as a political device to the EU and this situation disturbs the importer countries. Hence, the EU supports the power lines coming through Turkey since it wants to diversify the power resources. The Turkish Straits are the primary lines by which the power resources coming from Eurasian states are transported. It has transported 150 million tonnes of petroleum per annum. However, Turkey has started to create new transportation routes via the neighbouring countries because of the risks that could emerge during the transportation in the Straits. The current pipelines are Kirkuk–Yumurtalik, Baku–Tbilisi–Ceyhan, which has a transportation capacity of 50 million tonnes per annum, Baku–Supsa, also named Western Early, which has a transportation capacity of 16,000 tonnes per day, Baku–Novorossik, also named Northern Early, which has a transportation capacity of 14,000 tonnes per day, the Caspian Pipeline Consortium with 88,000 tonnes per day and finally Odessa–Brodi pipeline with 25,000–33,000 tonnes per day.[326]

The political crises between the EU and Russia increase Turkey's importance with each passing day. Russia uses the energy and natural gas in particular as a strategic policy device against the West. It is still remembered that Russia has been using natural gas as an oppression tool against Moldova, Georgia and Ukraine. This situation strengthens Turkey's hand over Russia in power security since the EU values the steady and reliable flow of the power above its price.[327]

The Middle East, the Caspian Region, Russia and Central Asia are the most prominent regions amongst which other energy resources are centered on. However, it is apparent that these power-rich countries have neither enough political reliability nor the geopolitical structure to open a big part of their power resources to the world markets. Thus, these countries are obliged to cooperate with other countries. Within this scope, Turkey plays an important role in the pipeline projects providing energy supply to Europe. The energy resources in the Caspian

---

326 **Kisacik, Şina:** Avrupa Birliği Enerji Politikasının Etkin Olarak Uygulanmasında Türkiye'nin Rolü, (Turkey's Role in EU Energy Policies) İstanbul, İKV Kütüphanesi, p. 18.
327 **İscan, İsmai l Hakkı (2007);** Türkiye-AB İlişkilerinin Geleceği Açısından AB Enerji Güvenliği Sorunu, (Turkey-EU Relations in the Framework of EU Energy Security) p. 159.

region were being transported to Europe via Russia. However, the EU's power security concerns and efforts to diversify its present supply motivated them to search for alternative routes in this field. Thus, Turkey has been conspicuous as the best and most reliable route because of both its geopolitical role and its position as candidate country to the European Union.[328]

Turkey has become roughly two thirds dependent on Russia thanks to the Blue Stream agreement. Besides, Turkey has been banned from marketing Russian gas coming from Blue Stream to the third world countries without Gazprom's permission and this situation has weakened Turkey's hand in this field. There are still ongoing arguments about this agreement in Turkish policy. It seems that this project militates in favor of Turkey in many aspects.[329]

Finally, there has been increasing cooperation in many fields as a result of Turkey's developing relations with the neighbouring countries in energy policy recently. Turkey is in a reliable position for both consumer and generator countries. Turkey's international and geopolitical position has gained strength by the policy developed with the neighbouring countries within this scope. Particularly, the TANAP-TAP project has increased Turkey's strategic status.

7.3 Interest and Cost of Cooperation on the Southern Gas Corridor

If we consider transnational relations, which lead to dependency and mutual dependence, according to Keohane and Nye, they are closely related with the development of international trade and transport relations. This kind of relationship and mutual dependence increases the dependency between countries and groups. Even totalitarian states allow their scientists and public officials to follow the developments of other states, although they are limited in some issues, which could be explained by mutual dependence. Dependency means that although the states know their behaviour is costly, the state has to follow certain policy. For instance, this issue has become integrated into the world monetary system and it makes it difficult for states to act autonomously as the state needs to provide capital. Foreign companies and technology, nationalist or socialist economic policies in underdeveloped states make it difficult to follow these policies. Even the state would need to change economic and financial policies, however, if international

---

328 **DELOITTE Consulting (2008)**; Energy Report of Turkey, p. 8.
329 **Demirmen, Fatih (2003)**; "Mavi Akım'da Neler Oluyor?" (What is Happening in Blue Stream?) Dünya Enerji Dergisi 35, pp. 40–42.

companies were damaged by their policies. These decisions can be made for the international companies.[330]

According to Keohane and Nye, the other effect of transnational relations is that some states give opportunity to others using such relations. Both sides are approximately equal to symmetric cases—for instance, the problems between the US and the Soviet Union; they are unequal countries in terms of transnational relations, which are categorised as asymmetric relations. Particularly within asymmetric relations, the advantageous position of a state has the ability to influence the other state. In this respect, even a tourist on a trip can act on the other state's ethnic and religious groups. Economic interaction by a government in terms of political and economical aims is used to realise objectives. Especially, strong states with quotas and tariff policies in line with their own interests use their power in order to influence international trade relations as well as using the less-developed countries. Furthermore, import of finished and semi-finished materials can prevent the development of production in these countries by placing high customs duties.[331]

In reference to Kant, who *"argued that interdependence strengthens the peaceful ties between states by creating incentives, economic ties provide a peaceful and cooperative international environment, and such incentives minimize the risk of international conflict. Thereby, an increase of economic interaction provides peace and diplomacy ties. According to liberalists and interdependence approaches, governments acquire more incentives by these interactions and seek protection by encouraging their governments to cooperate with other nations. On the other hand, interdependence assumption emphasizes that increased trade reduces the possibility of wars, which are far more costly than mutual cooperation."*[332]

From this perspective, international corporation could minimize the problem among states by projects such as the Southern Corridor. If the TANAP does not prove successful, the EU will never have a chance of getting diversification, resulting in weakness in foreign gas policy. On other hand, the South Stream is also a political project that will certainly be costly for Russia.

Considering Turkey's dependence on Russia for gas as well as oil and coal, Turkey has some doubts whether Russia could be used to impact Turkish foreign

---

330 **Arı, Tayyar (2011)**; International Relation Theory, Mkm yayincilik, Bursa, p. 401.
331 **Nye, Jr., Joseph S. /Robert O. Keohane (1972)**;, Transnational Relations and World Politics; An Introduction, Cambridge, Massachusetts Harvard University press, xix–xxi.
332 **Rosecrance, Richard (1986)**; The rise of the trading state commerce and conquest in the modern world, Basic Books, New York, pp. 102–103.

policy. In any case, Turkey received gas price discounts, which gave permission to construct the South Stream gas pipeline in the Turkish economic zone in the Black Sea. Some scientists criticised the permission for Turkey to construct the South Stream, as it can further damage the development of the Southern Gas Corridor.[333] In particular, in central and eastern Europe, the need to diversify the sources of natural gas imports exists because of problems manifested in the transit of Russian gas via Ukraine and lastly the Russo-Georgian War of August 2008. Particularly, these crises affect eastern and central Europe, which is mostly dependent on Russian gas imports.[334]

At this point, Ankara is dependent on gas from Iran and, mostly, Russia. This dependency results in vulnerability for Turkey. Therefore, dependency on Russian natural gas could be restricted through Turkey's foreign policy. Turkey does not have enough natural gas and crude oil reserves. In this sense, the AKP administration strived to explore other options by drilling in the Black Sea and the eastern Mediterranean. The result for Turkey has been disappointing because it will not commence in the near future.

On other hand, Turkey is becoming one of the world's fastest growing economies. The Turkish economy grew by over 3% in 2012 and it is expected to grow around 4% in 2013 and 5% in 2014. Energy demand will also increase correspondingly. In this regard, the AK arty administration is seeking to meet these needs to expand the use of renewable energy like wind, solar and geothermal power, and plans to build two more nuclear power plants in Turkey by 2023.[335]

Another important concern regarding gas is generated electricity in Turkey. Approximately 50% of electricity is from gas-fired power stations because it is considered more environmentally-friendly than oil-coal powered units and is also cheaper. Gas consumption in Turkey has been rising according to 2011 BP statistics. With reference to this rise in usage, Turkey imported 23 billion cubic meters from Russia through the Blue Stream pipeline.[336]

Finally, there has recently been increasing cooperation in many fields as a result of Turkey's developing relations with neighbouring countries in energy policy. Turkey is in a reliable position for both consumer and generator countries.

---

333 **Kardas, Saban;** Turkey-Russian Energy Relations "International Journal, Vol. 67. No. 1, Winter 2011-2012, pp. 97-98.
334 An EU Energy Security and Solidarity Action Plan- Second Strategic Energy Review (European Commission, November 13, 2008).
335 OECD Turkey - Economic Summary http://www.oecd.org/eco/outlook/turkeyeconomicforecastsummary.htm (accessed on 01/02/2013).
336 **EMRA (2012);** Natural Gas Market 2011 Sector Report.

Turkey's international and geopolitical position has gained strength by the policy it developed with the neighbouring countries within this scope. Particularly, the TANAP has increased Turkey's strategic status. The Commission addressed "*continuing cooperation with Turkey, aiming at integration of the country into the Energy Community.*"[337] Interdependence analyses results in the gas sector, especially on the level of vulnerability between Europe, most gas suppliers and transit countries and a slight asymmetry in favour of the EU. This situation can change in the gas cartel. Although a realisation of a common EU energy policy has not yet arisen, interdependencies between individual EU Member States still exist. These will manifest themselves when transnational projects such as the TANAP or the South Stream are implemented. The interdependencies occur on those political and economic levels, forcing the players to cooperate, meaning that various actors need each other to build a pipeline at a reasonable cost from Russia and the Caspian Sea to Europe.

7.4 Changes to Turkey's Position

As it was demonstrated in Chapter 5, the world gradually consumes more gas. Therefore, countries and regions demonstrate a global struggle for gas supply. As it was discussed in the study, the importance of gas will increase in the energy consumption of the next century. In today's world, energy dependency affects not only the economic but also the political decisions for the countries. The foreign dependency of countries in gas and oil causes political crises from time to time for the importing country. Therefore, these crises lead to a change in the decision authorities and foreign policies of these countries. The role of energy in Turkish foreign politics was not significant until the oil crisis of 1973. Until that period, Turkey had supplied Middle East oil and resources at cheap prices. With the increase of oil prices after the oil crisis, the economy experienced significant problems. Turkey's increasing dependency on the Middle East caused a shift from passive politics to active politics in foreign policy. Following the oil crisis, Turkey started to build economic and political bridges with Middle East and in particular European countries. The cooperation between Turkey and the European Economic Community is based on the Ankara Agreement, which was signed on 12 September 1963. The Customs Union Agreement was put in to force in 1996. Opposed to this, Turkey could only be accepted as a candidate country in 1997. At the end of the summit, it was stated that the country wished to become a full

---

337 **Council of European Union;** Council conclusions on strengthening the external dimension of the EU energy policy, Press release, Brussels, 24 November 2011.

member of the EU. In that period, Turkey started to implement Copenhagen criteria in economic and political fields. With the start of the negotiations between the EU and Turkey on 17 December 2004, in the report in which the European Commission mentioned the advantages to be provided with Turkey's membership to the EU, the role of Turkey in particular with regard to energy security was emphasised. Its strategic location and it being a transit country for the countries with gas and oil resources were effective in the commencement of the candidacy process. The increasing importance of Turkey in the field of energy had positive impacts on Turkey-EU relations and made it relatively easier to become full member. The energy lines reached Western Europe from countries such as Russia and Algeria following the routes outside Turkey. The importance of Turkey increased more after the year 2000. The BTC pipeline and the Azerbaijani natural gas pipeline, which was constructed after that, played a vital role in breaking Azerbaijan's dependence on Russia. As it was demonstrated in this study, if Kazakhstan, Iraq and Turkmenistan could follow the path that Azerbaijan did, this would have an impact on the policies of the region over time. There is no other country in the region that could play the role that Turkey plays in re-creating the balance of power in the region and the integration of the Turk Republics with the West through energy. The EU considers Turkey's close relations with its neighbouring countries positively from the point of view of developing alternative energy resources, as demonstrated by TANAP; Iran is the exception to this, as it is considered as problematic in the flow of energy from east to west. As it is well known, the pipelines bring countries together not only from an economic point of view, but also from a political point of view. In this way, the common areas of interest widen, communication becomes stronger and the perception of threat decreases. Turkey in particular adopts a significant target of integrating North Iraq with the region and the world through pipelines and other energy lines. Finally, Turkey hopes that its political importance will also be boosted as it turns into a hub where numerous energy lines intersect. As a conclusion, as it is claimed in this study, the role of gas will significantly increase within the context of global energy security in the 21st century, in particular between the EU and Turkey. Since the year 2000, energy has been playing a significant role in the EU–Turkey relationship. Projects such as BTC, BTE and finally TANAP have improved Turkey's strategic importance. As a result of the relationships carried out with Northern Iraq, negotiations are ongoing for transmitting the resources of the region to the West. In case of an agreement, an alternative will be created for the Russian gas and this corridor will change the balances in the region. The opening of Southern Gas Corridor is considered as a significant start with TANAP; it is also claimed that different projects will follow.

8. Conclusion

The dissertation claims that countries incur change in their foreign policies and political relationships in terms of energy security. What is the impact of energy and gas on the EU-Turkey relationship?

This study aimed to analyse the issue of natural gas and energy security in the context of EU-Turkish relations within the theoretical framework of interdependence theory, which was used as a complementary tool to give a comprehensive understanding and explanation of intergovernmental cooperation in international politics. Keohane and Nye's theory of interdependence is mentioned as well as their ideas on how asymmetrical interdependence can be a source of power and influence and predict international phenomena. In this study, these ideas are concerned with an analysis of the relations between the EU and Turkey in diversifying their natural gas needs via the Southern Gas Corridor.

After the Second World War, the world experienced a growth of economic exchange between nations. Thus, global trade volume has been increasing the last fifty years, and regional integration initiatives such as the European Union, the Asian Pacific Economic Cooperation (APEC), and the North American Free Trade Act (NAFTA) have been developed. In this regard, nations have increased economic activity and along with it, competition. The aims of nations are to increase wealth through economic exchange and to benefit by trading goods and services. Furthermore, world economic factors are changing as new oil and gas fields are discovered, or the implementation of nuclear power decreases the demand for oil and gas. Thereby, economic actors attempt to organise their relations with respect to the potential hazards that accompany their transactions. In this case, Keohane and Nye attempt to blend the wisdom of both realism and idealism by developing a coherent theoretical framework for the political analysis of interdependence, which was examined in the second chapter of this work in order to provide the reader with an overall picture of increasing mutual dependence. In general, mutual dependence is characterised as reciprocal effects between countries or actors in different countries. Furthermore, this theory advocates that transactions among states involve reciprocal costly effects, and this theory suggests that international cooperation could minimise the risk of international conflict by these interactions and reduce the possibility of wars. As Immanuel Kant emphasised, economic relationships provide peaceful and diplomatic ties.

As the study in the fourth chapter showed, world natural gas consumption is growing at an annual rate of nearly 2%. It is expected that the growing use of gas in the world will become a very strategic energy source for the future. OPEC and other institutions consider the potential scenarios in the case of gas disruption or damage. Crises affecting the rise in the prices of gas and oil include most recently the Libya revolution, the Arab Spring and the tsunami in Japan. In short, an interruption in the energy supply makes the world and its energy consumption needs especially sensitive to these kinds of crises. At this point, the world is interdependent in cases of crises.

In the third chapter, Turkey's role in the Southern Gas Corridor was analysed. It was argued that if the TANAP project is completed, the transported gas may carry costly effects to the EU and Turkey, as any crises or political uncertainties cause oil and gas prices to rise. According to Keohane and Nye, interdependence restricts state autonomy, which is a cost that comes with interdependent relationships, and it is argued that the benefits that both parties receive from this interdependence will outweigh the costs. Interdependence comes to play in the Southern Gas Corridor. After routes from Russia, Algeria and Norway, the Southern Gas Corridor would become the fourth major gas supply route to the EU, as it would be an important route to southern and central Europe. The opening of the Southern Gas Corridor is a key priority for the EU, since it will connect European gas customers with regions that possess long-term potential as important gas suppliers. It is expected to become a supply route for 10–20% of the EU's gas demand by 2020. It was argued that the diversity of gas supplies by TANAP may mitigate the risk of any supply disruptions. Turkey has to diversify its natural gas resources in case of an interruption by Russia and Iran.

As discussed in the sixth and seventh chapter of this study, the EU's energy dependence on natural gas has been progressively increasing and according to Eurostat 2009, the EU needs to import 64.2% of its natural gas, of which 34% is from Russia. Following Russia, Norway and Algeria are the other main exporters of natural gas. According to the EU Commission, it is estimated that the EU will import over 80% of its natural gas needs in 2030.

Turkish energy dependency on natural gas was analysed in the eighth chapter. Turkey has been one of the world's fastest growing economies in the last decade. The Turkish economy grew by over 3% in 2012 and is expected to grow around 4% in 2013 and 5% in 2014. Turkey's natural gas consumption was around 35.1 bcm in 2009 and is expected to reach 82.7 bcm by 2020. With regard to natural gas, Turkey is dependent on Russia for 57.9% of its needs and also on Iranian gas. Turkey imported 23 billion cubic meters from Russia via the Blue Stream pipeline. Fifty-seven percent of this natural gas is likely used in electricity production. The total rate of external dependency on natural gas is around 97.3%. In the framework of the aims of Turkey's gas policy, Ankara aims to reduce Turkey's maximum import share to below 50% by 2015. As for the power consumption, in parallel, it has increased eight times more. The deficit between the power generation and consumption has grown. A current account deficit exists due to the rise of the power share in the imports, particularly as Turkey has used natural gas for power generation in recent years.

In the ninth chapter the recent developments in the Southern Gas Corridor were examined. In order to diversify the natural gas of Europe and Turkey, states began the Nabucco project and made agreements within the states. However, the construction of the Nabucco pipeline was more expensive than estimated and Hungary had financial problems. Nabucco did not have enough gas reserves and Azerbaijan had no shares in Nabucco. For these reasons, the project was changed and revised as Nabucco-West and is shorter than Nabucco, which is from the Turkish-Bulgarian border to Austria, making it much cheaper than the original version. In this case, Azerbaijan and Turkey developed the natural gas pipeline TANAP, which is seen as an alternative Nabucco project. Shah Deniz Consortium will choose either Italy by TAP or the Nabucco-West project in June 2013. They are currently negotiating transportation tariffs and other commercial conditions. Shah Deniz Consortium is planning to extend extra gas from Turkmenistan as well as Iraq, potentially adding Kazakhstan and the Middle East in the future. Accordingly, it is evident that states are aiming to take part in an interdependent relationship based on the potential benefits and costs. The Consortium currently plans that if Turkmen gas is connected to TANAP, then TANAP's capacity could rise to 60 bcm, beginning in the first stage with 16 bcm of gas flowing. Moreover, Ankara is negotiating with KRG to allow natural gas from northern Iraq to be transported to Europe via Turkey. Thus, Turkey aims to supply 15 bcm to the Turkish market with additional volumes to Europe, and would link Iraq's natural gas to the Turkish national pipeline network. The gas reserves of the Kurdistan region of Iraq are estimated at 2.8-5-7 trillion cubic meters, which is about four times more than Azerbaijan's gas reserves. This source could become a game-changer in the long term with regard to the Southern Gas Corridor, and Turkey's position would be strengthened by this network. The only pipeline currently transporting small volumes of Azerbaijani gas is the Interconnector Turkey–Greece project. In the framework of the TANAP agreement, Ankara will get the right to transport additional gas to the Turkish market and get a rate for gas deliveries to Turkey.

As the study discussed, European energy security is seen by US administrations as a national interest. In this point, the US focuses on developing a southern corridor route for Caspian, Central Asian, and Middle Eastern natural gas to be transported by pipeline to Europe. George W. Bush criticised Russia for using energy supplies as a political weapon to affect other states and aimed to diversify European energy supply sources. The Obama administration also supports the diversification of the European gas supply. Due to past supply cutoffs, the future energy supply of the EU may be interrupted. Some EU states are looking for alternative routes supported by the United States. Otherwise, Germany decided to support the construction of the Nord Stream pipeline that directly connects Russia and Germany. At the same time, Russia has developed another pipeline project, the South Stream pipeline, which runs across the Black Sea. Besides bypassing transit states such as Ukraine and Belarus, it also bypasses EU states such as Poland and Lithuania. Through these proposal projects, Russia would be strengthened and in a position to dominate the European market share of natural gas. On this point, Russia tries to make bilateral agreements with major European countries such Germany, Italy and the Netherlands. However, some EU member states, such as Poland and Lithuania, opposed the construction of the Nord Stream pipeline. Russia's competition with the United States for influence over Caspian and Central Asian energy supplies was also analysed.

To summarise, because of Turkey's dependence on Russia for gas as well as oil and coal, construction of nuclear energy may become the next supply source for Turkey's energy needs. As mentioned above, Turkey is dependent on for natural gas, and it is argued that this dependency may restrict Turkey's foreign policy, as Iran escalated policy in winter and threatened to interrupt Turkey's natural gas supply. At the same time, the EU may be strongly dependent on Russian gas, through the planned South and Nord Stream pipelines that bypass Ukraine and Belarus. As it has been in the past, Russia may be used by domestic and foreign interests in the future. Ultimately, Europe desires security and diversification of supplies to decrease levels of vulnerability. Therefore, the EU must pay more attention to the proposed Southern Gas Corridor to diversify their natural gas imports. In short, it is argued that through international projects, international transactions minimise the problem within states and if the TANAP is not completed, the EU and Turkey may not achieve diversification, which would be seen as a weakness in the state's foreign gas policy. Furthermore, as a transit country, Turkey's interest and position may be harmed. On the other hand, the vulnerability of Europe is highly import-dependent on Russian gas, and the need to diversify sources of supply exists. In this regard, Keohane and Nye argue that the gas crises show the sensitivity aspect of interdependence. According to this, the EU and Turkey would, as actors, incur high costs from Russia, giving a reason as to why the policy should be changed to reduce these costs and risks. Russian–Ukrainian Projects,

Turkey and Blue Stream have an interdependent energy relationship which displays an asymmetrical interdependence.

Consequently, as discussed in the dissertation, Turkey wishes to strengthen its position as a new gas artery between the resource region and Europe. Turkey's status as a candidate country for EU membership in 2005 has been an important step in Turkey's development. Despite Greek–Cypriot opposition to opening the energy chapter, the integration and harmonisation period between Turkey and the EU has begun. In this regard, the opening of an energy chapter to negotiations may provide important momentum in the context of further alignment of the EU's internal gas market with Turkey. Furthermore, the EU and Turkey work together in the field of energy for the enhancement of energy security, such as reduction of pollution, trade of nuclear materials, creation of energy effiency programmes and the Kyoto Protocol by Turkey, as well as the initiation of several projects for companies such as Trans- European energy networks. In this case, Turkey decided to implement the European Commission Treaty in order to evaluate all possible outcomes of the ratification development in a legal framework. The power policy topic among the other topics of the negotiation period has a primary importance. It has been estimated that the cooperation between Turkey and the EU—including the planned project TANAP, which has been supported by the Commission—will make a significant contribution to Turkey's membership process, and can in turn be a catalyst for further and deeper cooperation between Turkey and the EU on energy matters in the future.

As seen, economics and political gains are significant parts of Turkey's accession to the EU. Accordingly, the interdependence approach argues that economic ties affect the political state's behaviour. Projects between states will aid in connecting their populations and strengthening political interactions among nations. Cooperation and economic interdependence on common projects will have effects on the political behaviour of states. Ultimately, these projects will accelerate Turkey's membership to the EU.

The study also gives an account of the global energy crises and their impact on the world as mentioned in Chapter 3. In particular, it was demonstrated that national policies were not as they had been in the aftermath of September 11, and that the international and regional powers are affected by each other in global crises. There is no country in the world that does not affect other countries in economic and political terms, as exemplified by the financial crises in Asia and the EU. The theory of interdependence claims that the countries of the world could work together to mitigate the impacts of the crises and that their political behaviours could change over time. As it was put by Keohane and Nye, international relations, state parameters and foreign policies could change over time. These relationships directly affect the foreign policy. In the aftermath of September 11, energy safety has been among the main agenda in the national and international arena. Oil prices continuously increased until 2003, and prices were doubled in 2007. Following the September 11 attack, the geopolitical and political tensions between countries increased remarkably. The Iraq war in 2003, the Arab Spring, the tsunami in Japan and lastly the revolution in Libya and the oil cutoff in 2011 all constituted a threat to global energy security. The gas crisis between Russia and Ukraine caused a hindrance in trade relationships between geopolitical actors after the 2006-2009 period. These crises demonstrate that foreign policy priorities change with energy policies and that countries thus became bound to one another in order to find alternative ways. With the increase of gas consumption, the need for gas increased and countries with gas reserves tried to find new resources to increase their production. As an example of this, the EU started to engage in relationships with Caspian countries and increased collaboration for the opening of the South Gas Corridor.

Critique; as expected, the EU must be able to make joint and effective decisions in foreign policy. Accordingly, it is contemplated that there are various reasons for the slow progress in the foreign policy mechanisms of the EU. As an example the recent EU–Ukraine crisis can be considered to see the effectiveness of the EU's foreign policy system. First of all, EU decision-making mechanisms fail to act rapidly in foreign policy issues, which require approval of all members. Similar problems are also encountered in the joint energy policy. Since member countries fail to develop a joint energy policy, each country arranges different bargaining agreements with Russia in energy issues and apply different conditions.

Due to the fact that EU countries are excessively dependent on Russia for energy, it is thought that this weakens the strength of the reaction against the Crimea crisis and that the foreign policy is shaped accordingly. In particular it is considered that actions are taken slowly against the crisis due to the probable damage, which is caused by any possible interruption of gas supply from Russia to the economy, in particular for the main EU countries such as the UK and Germany. Besides, it is also possible to state that this crisis reinforced Turkey's position as a country through which the energy lines pass. For that reason, Turkey should encourage preparation of relevant projects in order to ensure that its energy sources in Caucasia and Central Asia pass through its own territories.

As it is known, energy supply security is essential and affected by regional and natural factors and political conflicts. In particular, uncertainty in energy supply security can result in an increase in oil and energy prices. For that reason, EU member countries are required to keep 90-day oil and gas reserves against any possible interruption. Taking into account the fact that the EU's energy dependence will increase in the coming 20–30 years, the foreign policy mechanisms of the EU become more vulnerable against the Russian politics due to the lack of a joint policy. In particular, the differences between member countries should be abolished and the foreign policy should be structured according to EU norms rather than regional politics.

Political cooperation should occur as a result of EU trade relations which are designed within the context of the Common Market, customs union and being a regional commercial block. One of the main targets of the EU is to increase competition between member countries and keep pace with the USA and Japan. In addition to this, as it was indicated in the energy group of the European Parliament, despite the indication that the EU needs a robust joint energy policy, steps were not taken at the required level. In order to ensure the security of gas supply in EU countries, the Southern Gas Corridor is the most concrete and significant project. Although there was no high amount of gas that flew from the pipeline in question, it is expected that this will be developed with different projects.

On the other hand, the political desires could not be seen in the Nabucco project. Political stability could not be ensured due to the diverse energy policies of the member countries. It is recorded in history as an unsuccessful project with no result, though being characterised as the EU's project of the century. The TANAP project is considered as an alternative project. Different from the Nabucco project, the TANAP project does not comprise Western EU countries. As it is known, the dependence of these countries on Russia is ongoing. Although it is envisaged to expand the line with the Southern Gas Corridor project in the long run, it seems that it will take time to diversify energy supply.

With the Blue Stream natural gas agreement, Turkey's dependence on Russia increased within the context of gas supply. With the approval of this project, procurement of gas from Central Asia will be terminated and also the transfer of Central Asian gas to Europe through Turkey will be prevented. Meanwhile, it has tied its coming 25 years with the Blue Stream project. As a conclusion, the Blue Stream caused a delay for the Shah Sea project and the Trans Khazar project has been shelved. Besides, whereas Turkey could have purchased Turkmen natural gas at a very cheap price with the Trans Khazar project, with this agreement concluded it has signed many non-economic projects by preferring the same natural gas over Russia's.

As a consequence, the increasing gas consumption of the EU and Turkey forced the regions to collaborate with one another. The fact that both Turkey and the

EU are dependent on gas from abroad causes a threat for political and economic situations. These crises were elaborated upon in this study and it was claimed that the role of gas will prompt an increase in global energy security in the 21st century, also increasing the projects that will lead to political unions over time. Joint energy projects such as BTC, BTE, the Turkey–Italy–Greece gas pipeline, the Kirkuk–Ceyhan oil pipeline and finally TANAP lead to collaboration, and bring the respective societies together, forging diplomatic relations. In this regard, with the opening of the Southern Gas Corridor, interdependency between the EU and Turkey will increase. Other pipeline projects will accelerate with the opening of Shah Deniz I and II gas to the EU market. In addition to the opening of the Southern Gas Corridor, with its improved relations with the Northern Iraq regional administration, Turkey's position as a transit country for oil and gas resources to the EU countries will also improve.

9. References

Amy Myers Jaffe/Ronald Soligo (2004); Re-evaluating US Strategic Priorities in the Caspian Region: Balancing Energy Resource Initiatives with Terrorism Containment, Cambridge Review of International Affairs, Cilt 17, Sayı 2.

Ari, Tayyar (2001); Die internationale Beziehungen und die Theorien, Mkm yayincilik, Bursa.

Ari, Tayyar (2004); Uluslararası İlişkiler ve Dış Politika, Alfa yayincilik, Istanbul.

Ali, Yigit (2001); Elektrik Enerjisi Politikaları, Doğalgaz & Enerji Yönetimi Kongre ve Sergisi, Gaziantep.

Aad, Correlje/Coby, van der Linde (2006); Energy Supply Security and Geopolitics: A European Perspective, Energy Policy, Volume 34, 2006, Issue 5 March, pp. 532–543.

Alemdaroglu, Nusret (2007); Enerji Sektörünün Gelecegi Alternatif Enerji Kaynakları ve Türkiye'nin Önündeki Fırsatlar. İstanbul Ticaret Odası Yayınları, İstanbul.

Adellman, Moris (1993); The Economics of Petroleum Supply: 1962–1993, Cambridge, The Mit Press.

Andrews Speed, Philip (2003); Energy Security in East Asia: a European View. Presented at the Symposium on Pacific Energy Cooperation 12–13 February, Tokyo.

Andrew, Monagham (2006); Russia–EU Relations: An Emerging Energy Security Dilemma, Pro et Contra, vol. 10, issue 2–3 (Summer 2006), Carnegie Moscow Center.

Andersen, Svein S. (2000); European Integration and the Changing Paradigm of Energy Policy: The Case of Natural Gas Liberalisation, Arena Working Papers,13/2000.

Aras, Bülent/Yorkan, Arzu (2005); Avrupa Birliği ve Enerji Güvenliği: Siyaset, Ekonomi ve Çevre (TASAM Stratejik Rapor), No:13, Ankara, Tasam Yayınları.

Ariel, Cohen (2007); Europe's Strategic Dependence on Russian Energy, The Heritage Foundation, No. 2083, November 2007.

Aydin, Mustafa (2005); Türkiye nin Orta Asya–Kafkaslar Politikası, Nobel Yayinlari, Ankara.

Anar, Hüseynov (2002); Gas Perspectives of Azerbaijan, Economic News Bulletin, No.3.

Blank, Stephen (2008); The Strategic Importance of Central Asia: An American View, Parameters, Spring, Band 38 Nr. 1, April 2008.

Brown, Seyom (1996); International Relations in a Changing Global System: Toward a Theory of the World Polity, Westview Press.

Baklacı, Pinar/Akıntürk, Esen (2006); Enerji Şartı Anlaşması, Dokuz Eylül Üniversitesi İşletme Fakültesi Dergisi, Cilt. 7, Sayı.2.

Belkin, Paul, (2008); Report for Congress The European Union's Energy Security Challenges Updated January Analyst in European Affairs Foreign Affairs, Defense, and Trade Division.

Birol, Fatih (2007); World Energy Prospects and Challenges. Asia-Pacific Review, Vol. 14, No.1

Bacik, Gökhan (2006); Turkey and Pipeline Politics, Turkish Studies, Vol. 7, No.2.

Christiansen, Thomas/Jorgensen Knud Erik/Wiener, Antje (1999); The social construction of Europe, Journal of European Public Policy 6, no. 4, p. 529, cited in Özcan, Mesut (2008); Harmonizing foreign policy.

Chevalier, Jean Marie (2005); Security of Energy Supply for the European Union, "European Review of Energy Markets", Vol. 1, Issue 3 November.

Chapmann D./ Khanna N. (2006); The Persian Gulf, Global Oil Resources, and International Security, Contemporary Economic Policy, Vol. 24, Issue 4, pp. 507–519 October.

Copy, Van der Linde (2008); Turning a weakness into a strength, a smart external energy policy for Europe. Clingendael International Energy.

Cooper, Richard (1968); The Economics of Interdependence in the Atlantic Community, Mc Graw Hill Company, New York.

Canbolat, İbrahim (1998); Uluslar Üstü Sistem AB Bir Dönüşümün Analizi, İkinci Basım, İstanbul: Alfa Basım Yayım.

Davutoglu, Ahmet (2008); Turkey s Foreign Policy Vision: An Assessment of 2007; Insight Turkey,Vol. 10, No.2008.

Davutoglu, Ahmet (2001); Strateji Derinlik, Türkiye, nin Uluslararasi Konumu, Küre yayinlari.

Demirbas, Ayhan (2003); Fuel wood Characteristics of Olive Huskand Walout, Sunflower and Almound Shells, Energy Sources, No 25.

Dura, Cihan/Atik, Hariye (2003); Avrupa Birliği Gümrük Birliği ve Türkiye, Ankara: Nobel Yayınları.

Demirmen, Ferruh (2003); Mavi Akım'da Neler Oluyor? Dünya Enerji Dergisi, Sayi 35, Eylül.

Daniel, Yergin (2005); Energy Security and Markets, in *Energy and Security: Toward a New Foreign Policy Strategy*, J. Kalicki and D. Goldwyn, Editors, 2005, Woodrow Wilson Center Press: Washington, DC.

Erdogan, Murat, (2011); Middle East Policy of European Union, Report, Ankara.

Eray, Aynur (2002); Enerjide Tutumluluk ve Verimlilik, Temiz Enerji Vakfı Yayınları, İstanbul.

Ege, Yavuz (2004); Avrupa Birliği'nin Enerji Politikası ve Türkiye'nin Uyumu, 1.B., Uluslar Arası Politika Araştırmalar Vakfı, Ankara.

Ergun, Çağdaş Evrim (2007); Avrupa Birliği Enerji Hukuku, Caköak yayinevi, Ankara.

Energy Information Administration (2007c); Caspian Oil and Gas Production and Prospects.

EPDK:(2011); Turkish Energy Market: An Investor's Guide.

ETKB, (2011); Turkish Energy Policy, Ankara.

European Commission (2008g); Baku Initiative.

Energy and Natural Resources Ministry (2012); 2010–2014 Strategic Planning.

European Commission (2011); Energy 2020A strategy for competitive, sustainable and secure energy.

European Commission (2012); Turkey–EU Positive Agenda Enhanced EU–Turkey Energy Cooperation outcome of meetings.

European Commission (2001); Proposal for a directive of the European Parliament and of the Council establishing a framework for greenhouse emission trading within the European Community and amending Council Directive 96/61/EC.

European Commission (2000); Towards a European strategy for the security of energy supply.

European Commission, Green Paper (2001); Towards a European strategy for the security of energy supply.

European Commission: Treaty establishing the European Economic Community, EEC Treaty.

European Commission (2010); EU-Russia Energy Dialogue, Brussels.

European Commission; Summaries of EU Legislation, Treaty establishing the European Coal and Steel Community, ECSC.

European Commission (2008b); Climate Change and International Security, Paper from the High Representative and the European Commission to the European Council.

European Commission (2008h); EU's relations with Central Asia, Brussels.

Energy in Europe (1996); European Energy to 2020, European Commission Directorate General for Energy Brussels, Special Issue.

Energy Charter Treaty (1994); A Reader's Guide, Brussels.

European Commission (2010); Energy, transport and environment indicators, Eurostat.

European Commission (2011); Energy, transport and environment indicators, 2011 edition, Eurostat.

Featherstone, Kevin/ Kazamias, George (2011); Europeanization and Southern Periphery, Frank Cass, London.

Finon, Dominique / Locatelli, Catherine (2007); Russian and European gas interdependence. Can market forces balance out geopolitics?, Laboratoire d'Economie de la Production et de l'Intégration Internationale, Cahier de Recherche LEPII, Série EPE, Nr. 41, Grenoble 2007

Foreign Economic Council (2006); Energy Review.

Steinvorth, Daniel (2005); Friedrich Ebert Stiftung- Frankreich- Info Nr. 3 (2005 (Deutsche–französische Energiepolitik im europäischen Kontext)

Frederick, Starr (Ed.) Svante, Cornell (Ed.) (2005); The Baku-Tbilisi-Ceyhan Pipeline: Oil Window to the West.

Green Paper (2001); Towards a European Strategy for the Security of Energy Supply, Brussels European Commission.

Green Paper, (2006); A European Strategy for Sustainable, Competitive and Secure Energy, Brussels, Commission of the European Communities.

Gawdat, Bahgat (2006); Europe's Energy Security: Challenges and Opportunities, International Affairs, London, 82:961–975 September.

Graf Lambddorff, Alexander/ Chatzimarkakis, Jorgo (2008); Für die Gruppe der Europäischen Parlament Europäische Energiepolitik, Positionpapier.

Götzö, Roland (2007); Die russisch zentralasiatische Energiegemeinschaft – eine Bedrohung für die europäische Energiesicherheit, Diskussionpapier.

Graham E. Fuller (2008); The New Turkish Republic, Washington DC: United States Institute of Peace Press.

Heather, Grabbe (2003); Europeanization Goes East: Power and Uncertainty in the EU Accession Process, in; Featherstone, Kevin/ Radaelli, Claudio M. (Ed.) The Politics of Europeanization, Oxford University Press, Oxford.

Helm, Dieter (2005a); The Assessment: The New Energy Paradigm. Oxford Review of Economic Policy vol.21 No.1, Oxford University Press.

Hope, Kerin (2006); "Rice Bid to Exclude Russian Supplier", Financial Times, April 25.

Dieter, Helm (2005b); European Energy Policy: Securing Supplies and Meeting the Challenge of Climate Change, New College, Oxford, 25 October

Hepbasli, Arif /Özdamar, Aydogan/Özalp, Nesrin (2001); Present Status and Potential of Renewable Energy Sources in Turkey, Energy Sources, Taylor & Francis.

Horsnell, Paul (2004); Why Oil Prices Have Moved Higher, Oxford Energy Forum, August.

Hacisalihoglu, Bilge, (2008); Turkey's natural gas policy: Energy Policy, June:1870.

Jonathan, Stern (1987); Soviet Oil and Gas Exports to the West: commercial transaction or security threat?, Gower Pub. Co. Energy Policy Studies.

Kibaroglu, Mustafa (2004); Dünya ve Türkiye deki Enerji ve Su Kaynaklarının Ulusal ve Uluslararası Güvenliğe Etkileri, Harp Akademileri Komutanlığı Silahlı Kuvvetler Akademisi, İstanbul.

Kröger, Wolfgang (2006); Issues of Secure Energy Supply. Symposium, October 12, Zürich.

Keskin, Hakan (2007); Genişleme Ve Derinleşme Süreçlerinde Avrupa BirliğiPolitikaları, Stratejik Araştırmalar Dergisi, Genelkurmay Askeri Tarih Ve Stratejik Etüt Başkanlığı Yayınları, Sayı. 9, Yıl.5.

Keskın, M. Hakan; Stratejik Açıdan Avrupa Birliği Enerji Politikası ve Uluslararası Güvenlik Sistemine Etkisi, Dokuz Eylül Üniversitesi Sosyal Bilimler Enstitüsü Avrupa Birliği Anabilim Dalı Yayımlanmamış Doktora Tezi

Kardas, Saban; Turkey–Russian Energy Relations International Journal, Vol. 67, No. 1, Winter 2011–2012.

Krasner, Stephen (1983); International regimes, Cornell University Press, from Krasner's overview essay.

Karen, Mingst (2004); Essential of International Relations, W.W. Norton, third edition.

Keppler, Jan Horst (2007); International Relations and Security of Energy Supply: Risks to Continuity and Geopolitical Risks, European Parliament's Committee on Foreign Affairs, Brussels.

Keohane, O. Robert/S. Nye, Joseph (1989); Power and Interdependence, Harper Collins Publishers.

Keohane, O. Robert/S. Nye, Joseph (2001); Power and Interdependence, 3rd edition Longman.

Keohane, O. Robert/S. Nye, Joseph (1977); Power and Interdependence: World Politics in Transition, Brown Company, Boston.

Keohane, Robert O./Nye, Joseph (1987); Power and Interdependence: Revisited, International Organization, Vol. 14, No.4.

Kucera, Joshua (2012); Armenian Military Simulates Attack on Azerbaijan's Oil.

Katharina, Graisy (2009); Die Interdependenz zwischen Russland und der Ukraine unter Berücksichtigung der Ressource Erdgas, Diplomarbeit/Universität Wien.

Kücüksahin, Ahmet (2006); Türkiye'nin Enerji Stratejisi Ne Olmalıdır?, Genel kurmay Başkanlığı Yayınları.

Kisacik, Sina; Avrupa Birliği Enerji Politikasının Etkin Olarak Uygulanmasında Türkiye'nin Rolü, İstanbul, İKV Kütüphanesi

Kramer Heinz (2010); Die Türkei als Energiedrehscheibe, Wunschtraum und Wirklichkeit, SWP- Studien.

Iscan, Ismail Hakki, (2007); Türkiye-AB İlişkilerinin Geleceği Açısından AB Enerji Güvenliği Sorunu Uluslararası Ekonomi ve Dış Ticaret Politikaları Dergisi.

Inogate Programme, (2008a); Energy Cooperation between the EU, Eastern Europe, the Caucasus and Central Asia

Ian, Bartle: (1999); Transnational Interests in the European Union: Globalization and Changing Organization in Telecommunications and Electricity, Journal of Common Market Studies, Vol. 37, issue 3.

Iktisadi Kalkınma Vakfı (2004); Avrupa Birligi'nin Enerji ve Ulastırma Politikaları ve Türkiye'nin Uyumu, Istanbul

International Energy Agency (2003); Energy Policies of IEA countries, 2003 review.

Julian, Horn Smith (2008); Turkey: Trade and EU Accession, Adam Hug (ed.), Turkey in Europe: The Economic Case for Turkish Membership of the European Union, London, The Foreign Policy Center.

Joint Press Release, Turkey and the EU (2007); Together for a European Energy Policy High Level Conference.

Leite, Carlos/Weidmann, Jens (1999); Does Mother Nature Corrupt? Natural Resources, Corruption and Economic Growth, IMF Working Paper, WP/99/85 Washington DC.

Lavenex, Sandra/Schimmelfennig Frank (2007); Relations with the Wider Europe, Journal of Common Market Studies Annual Review of the European Union Volume 45, Issue 1.

Lehmkuhl, Dr. Ursula (2001); Lehr und Handbücher der Politikwissenschaft, Lehrmkuhl, Theorien Internationaler Politik, Oldenburg, 3. Auflage.

Lemke, Christiane (2001); Internationale Beziehungen Grundkonzepte, Theorien und Problemfelder, 2. Auflage München.

Larrson, R. L. (2008); Europe and Caspian Energy: Dodging Russia, Tackling China and Engaging the U.S. Europe's Energy Security: Gazprom's Dominance and Caspian Supply Alternatives, (19-41) Derleyen Svante Cornell & Niklas Caucasus Washington: Institute & Silk Road Studies Program.

Laciner, Sedat/Özcan Mehmet/Bal, İhsan (2005); European Union With Turkey: The Possible Impact Of Turkey's Membership On The European Union, Ankara, USAK.

Meyers, Reinhard (1979); Weltpolitik in Grundbegriffen, Ein lehr und ideengeschichtlicher Grundriss, Droste Verlag.

Morse, Edward Lewis (2002); The Battle For Energy Dominance, in: Foreign Affairs (March/April).

Morse, Edward L. (1972); Transnational Economic Processes, Keohane and Nye Transnational Relations and World Politics, Massachusetts, Harvard University Press.

Moravcsik, Andrew (1993); Preferences and Power in the European Community: a Liberal Intergovernmentalist Approach; Journal of Common Market Studies.

Milovzorov, Andrei (2006); "Gazprom navodit uzhas na Evropu", *Ekonomika*, March 3, 2006.

Müsiad (2006); Türkiye'nin Enerji Ekonomisi ve Petrolün Gelecegi. Arastirma Yayınları, Istanbul.

Nye, Jr., Joseph S./Robert O. Keohane (1972); Transnational Relations and World Politics; An Introduction, Cambridge, Massachusetts Harvard University press.

Nye, Joseph (2005); Soft Power: The Means to Success in World Politics, public Affair.

NATIONAL PROGRAMME OF TURKEY for the Adoption of the EU Acquis (2001-2003)

Oran, Baskın (Ed.) (2006); Türk Dış Politikası II, 1980-2001 (Turkish Foreign Policy II 1980-2001) Ilhan Uzgel, ABD ve Nato, yla Iliskiler (Relations with USA and NATO), İletişim, İstanbul.

Oran, Baskin (2010); Türk Dis Politikasi Cilt II: 1980-2001 Iletisim Yayinlari (E.d) Mustafa Aydin (Kafkasya ve Orta Asyayla Iliskiler), Iletisim, Istanbul.

Oye, Kenneth (1986); Explaining Cooperation Under Anarchy, Princeton University Press.

O'Rourke, Breffni (2006); Caspian: EU Invests In New Pipeline. Radio Free Europe / Radio Liberty.

Özkan, Gökhan (2010); The Energy Security Dimension Of Turkey's Regional Policy In Central Asia And The Caucasus, Akademik Bakış Cilt 4 Sayı 7 Kış, 2010.

Özcan, Mesut (2008); Harmonizing foreign policy, Turkey, the EU and the Middle East, Ashgate, Aldershot.

Ostry, Sylvia (2006); Sustainable Development and Energy Security: The WTO and the Energy Charter Treaty, Paper, Moscow.

Özer, Serra (2004); A Feasibility Study and Evaluation of Financing Models for Wind Energy Projects: A Case Study on Izmir Institute, A Dissertation, Izmir Teknoloji Üniversitesi.

Oktay, E./R. F. Cankıran (2006); Avrupa Birliği'nin Enerji Güvenliği Açısından Türkiye'nin Önemi, Avrupa Araştırmaları Dergisi, Cilt:14, Sayı.1, Istanbul.

Pascual C./Zambetakis E. (2010); The Geopolitics Of Energy From Security To Surviva, in: Energy Security: Economics, Politics, Strategies, and Implications editied (C. Pascual and J. Elkind), Washington; The Brookings Institution,9–36.

Petersen, Alexandros (2007); Integrating Azerbaijan, Georgia, and Turkey with the West: The Case of the East-West Transport Corridor, Center for Strategic and International Studies.Working Paper, October 2.

Pamir, Necdet, (2005); AB'nin Enerji Sorunsalı ve Türkiye Stratejik Analiz Dergisi, Cilt.6, Sayı.67.

Pala, Cenk. (2005); Hazar Bölgesi ve Türkiye Açısından Önemi Konferans, Working Paper.

Roberts, John, (2004); The Turkish Gate Energy Transit and Security Issues, CEPS EU-Turkey Working Papers, No. 11.

Robbert, Willenborg/Christoph, Tönjes/Wilbur, Perlot (2004); Europe's Oil Defences: An analysis of Europes's oil supply vulnerability and its emergency oil stockholding systems, The Clingendael Institute, The Hague.

Rosecrance, Richard (1986); The rise of the trading state commerce and conquest in the modern world, Basic Books, New York.

Roberts, John (2004); The Turkish Gate Energy Transit and Security Issues, CEPS EU-Turkey Working Papers, No. 11.

Stubbs, Richaerd/Underfhill, Geoffrey (2006); Political Economy and the Changing Global Order:2006, Oxford University Press.

Spelmenn, Scott (2007); The NATO School Energy Security Conference, Issue Paper Nr. 3.

Spindler, Manuela (2006); Interdependenz, in: Schieder, Siegfried/Spindler, Manuela (Hrsg.), Theorien der Internationalen Beziehungen, 2. Auflage, Opladen.

Siegfried, Schieder/Manuela Spindler (2006); Theorien der Internationalen Beziehungen (Hrsg.) Opladen Hill.

Sullivan, Michael (1989); Transnationalism, Power Politics, and the Realities of the Present System, International Relations in the Twentieth Century, A Reader, Marc Williams (E.d) London: Mach Millan Education.

Schimmelfenning, Frank (2008); Internationale Politik. München, Wien, Zürich, UTB Schöningh Verlag.

Socor, Vladimir (2012); Eurasia Daily Monitor, Volume: 9, Issue: 148, August 3, 2012, Azerbaijan-Europe Gas Transportation Consortiums Face Major Restructuring.

Saygın, Hasan (2004); Sürdürülebilir Gelisme Gündeminde Nükleer Enerjinin Sorunları, Elektrik Mühendisligi Dergisi.

Sodupe, Kepa/Benito, Eduardo (2001); Pan-European Energy Co Operation Opportunities, Limitations, and Security of Supply to the EU, Journal of Common Market Studies, Vol. 39, Issue 1, March.

Soylu, Hakkı, (2007); Rusya, Bulgaristan ve Yunanistan'ın Üzerinde Anlaştığı Burgaz-Dedeağaç Petrol Boru Hattı'nın Türkiye Açısından Önemi, Silahlı Kuvvetler Dergisi, Sayı.391.

State Planing Organisation (2004); Secretariat General for EU Affairs, National Program of Turkey's undertaking of the Acquis of the EU Ankara, October 2004.

State Planing Organisation (2003); Secretariat General for EU Affairs, National Program of Turkey's undertaking of the Acquits of the EU EU Progress Report of 2003 Ankara.

Stern, Jonathan (2006); Security of European Natural Gas Supplies: The Impact of Import Dependence and Liberalization, The Royal institute of International Affairs, London.

Turkey Energy Strategy (2009); Republic of Turkey Ministry of Foreign Affairs, Energy, Water and Environment General Director of Turkey.

The Republic of Turkey, Secretariat General for EU Affairs (2003); ABGS, National Program of Turkey's Undertaking of the Acquits of the EU.

Tönjes, Christoph / Jong, Jacques (2007); Perspective on Security of Supply in European Natural Gas Markets, Clingendael International Energy Programme Working Paper.

Tüsiad (2003); Türkiye'nin Enerji Sorunları ve Çözüm Önerileri.

Tüsiad (Turkish Industry and Business Association) (2007); Avrupa Birliğine Katılım Sürecinde Türkiye'nin Komsu ve Çevre Ülkeler Politikası.

The Republic of Turkey, Secretariat General for EU Affairs (2008); ABGS, National Programme of Turkey for the Adoption of the EU Acquis.

Tonus, Özgür (2004); Genişleyen AB'nin Enerji Politikaları ve Türkiye, Müzakere Sürecinde Türkiye AB İlişkileri Uluslararası Sempozyumu. Ankara: Gazi Üniversitesi.

Torbjorn, L. Knutsen (1992); History of International Relations Theory, Manchester University Press.

Umbach, Frank (2003); Globale Energiesicherheit, Oldenbourg Verlag, München.

United States Energy Association (2002); National Energy Security Post 9/11 Washington, D.C.: United States Energy Association.

Ültanir, Mustafa (1998); 21. Yüzyıla Girerken Türkiye'nin Enerji Stratejisinin Degerlendirilmesi, Lebib Yalkım Yayınları, Istanbul.

Wiener, Antje/Diez, Thomas (2004); European Integration Theory, Oxford University Press, Oxford, cited in Özcan, Mesut (2008); Harmonizing foreign policy, Turkey, the EU and the Middle East, Ashgate, Aldershot.

Westphal, Kirsten (2006); Energy Policy between Multilateral Governance and Geopolitics: Whither Europe. Journal of International Politics of Society.

Waltz, Kenneth (1979); Theory of International Politics, Addison-Wesley Pub. Co.

William, Drozdia (2006); Russia: More Awkward, But Still Indispensable, European Affairs, Vol. 7 Issue 1–2 in the Spring/Summer of 2006.

Woehrel, Seven (2008); Russian Energy Policy Toward Neighboring Countries. CRS, Report for Congress, November 27.

Vitale, Alessandro (2007); The EU Wants to Build An Energy Strategy in The Caspian Regionö Caucaz Europe News, 01/07/2007.

Vander, Linde. C (2002); The State and International Oil Market, Competition and the Changing Ownership of Crude Oil Asset. Boston/Dordrecht/London, Kluwer Academic Publishers.

Yavuz, Ege (2004); Avrupa Birliği'nin Enerji Politikası ve Türkiye'nin Uyumu, 1.B., Ankara, UPAV.

Yenal, Kesbic, Cüneyt/Simsek, Hasan (2001); Avrupa Birliği Ortak Enerji Politikası, Muğla Üniversitesi SBE Dergisi, Sayı 5.

Yığıt, Ali (2001); Elektrik Enerjisi Politikaları, Doğalgaz & Enerji Yönetimi Kongre ve Sergisi, Gaziantep.

INTERNET

http://www.bilgesam.com/tr/index.php?option=com_content&view=article&id=371:kuresel-enerji-denklemindeturkiye&catid=131:enerji&Itemid=146

http://www.haber3.com/botas-enerji-sektorunde-buyuk-oynayacak-745098h.htm

http://zaman.com.tr/yazar.do?yazino=1209175

http://www.botas.gov.tr/index.asp BOTAS Petroleum Pipeline Corporation

http://botasint.com/SirketProfili.aspx BOTAS International Limited

http://www.bp.com/sectiongenericarticle.do?categoryId=9006669&contentId=7015093 Bakü–Tiblisi–Ceyhan Pipeline

http://www.calikenerji.com/eng/rafineri.php?sf=rafineri_yatirim&k=283&u=422 Samsun–Ceyhan Crude Oil Pipeline

http://www.enerji.gov.tr/index.php?dil=en&sf=webpages&b=enerji_EN&bn=215& hn=&nm=40717&id=40717 Republic of Turkey Ministry of Energy and Natural Resources

http://gazprom.com/production/projects/pipelines/blue-stream/ Gazprom: Blue Stream project between Turkey and Russia

http://ei-01.eia.doe.gov/countries/cab.cfm?fips=RS Energy Information Administration: 2009, Russian Country Analysis Briefs

http://south-stream.info/index.php?id=9&L=1 Southstream official Page: Europe's Energy Security

http://www.heritage.org/research/reports/2007/11/europes-strategic-dependence-on-russian-energy

http://portal.nabuccopipeline.com/portal/page/portal/en/company_main/about_us Nabucco Gas Pipeline Company

http://www.enerji.gov.tr/index.php?dil=tr&sf=webpages&b=dogalgaz&bn=221&hn=&nm=384&id=40694 Republic of Turkey Ministry of Energy and Natural Resources: Natural Gas

http://www.haber7.com/haber/20100607/Azerbaycanla-Sahdeniz-projesi-imzalandi.php- Shah Deniz Project

http://enerjienstitusu.com/2011/06/15/musiad-nukleer-enerji-zaruri zorunluluk/#ixzz1PNbNLjRq http://www.world-nuclear.org/info/inf102.html World Nuclear Energy

http://www.haberler.com/basbakan-erdogan-enerji-dagiliminda-adalet-2258967- haberi/

http://www.euractiv.com/energy/nabucco-pie-sky-georgia-crisis/article-174855

http://www.bilgesam.com/tr/index.php?option=com_content&view=article&id=371:kuresel-enerji-denklemindeturkiye&catid=131:enerji&Itemid=146

http://www.euractiv.com/en/energy/coal-clean-energy-source-future/article-156397 Euractiv

http://www.encharter.org/index.php?id=37

Energy Framework Programme ETAP http://ec.europa.eu/energy/evaluations/doc/2002_energy_framework_progra m.pdf

The Carnot Programme http://ec.europa.eu/energy/evaluations/doc/2002_energy_framework_program.pdf

The Save Programme http://ec.europa.eu/energy/evaluations/doc/2002_energy_framework_program.pdf

The Altener Programme http://ec.europa.eu/energy/evaluations/doc/2002_energy_framework_program.pdf

Energy Framework Programme SURE http://ec.europa.eu/energy/evaluations/doc/2002_energy_framework_program.pdf

Energy Framework Programme SYNERGY" http://ec.europa.eu/energy/evaluations/doc/2002_energy_framework_program.pdf

"White Paper: An Energy Policy for the European Union" http://europa.eu/documents/comm/white_papers/index_en.htm

European Commission, Environment: The Sixth Environment Action Programme 2002–2012 http://eurlex.europa.eu/LexUriServ/LexUriServ.do?uri=OJ:L:2002:242:0001:0015: EN:PDF http://ec.europa.eu/environment/newprg/intro.htm

Energiesicherheitspolitikistauch Friedenspolitik. Walter Steinmeier, am 16.2.2007, www.auswaertiges-amt.de/diplo/de/Infoservice/Presse/Reden/2007/070216 Energiekonferenz.htm

EU-Russia Center, "The EU-Russia Center Review: EU Russia Energy Relations", Issue 9, June 2009, retrieved from http://www.eurussiacentre.org/wpcontent/uploads/2008/10/review_ix.pdf

http://diepresse.com/home/wirtschaft/economist

http://www.hurriyetdailynews.com/russia-starts-16-bln-euro-gasproject.aspx?pageID=238&nID=36403&NewsCatID=348

http://www.eurasianet.org/node/66061

http://www.haber3.com/botas-enerji-sektorunde-buyuk-oynayacak-745098h. htm

http://www.enerji.gov.tr/index.php?dil=tr&sf=webpages&b=enerji_cevre_iklim&bn=218&hn=&id=4303

http://enerjienstitusu.com/2011/06/15/musiad-nukleer-enerji-zaruri zorunluluk/#ixzz1PNbNLjRq

http://www.oecd.org/eco/outlook/turkeyeconomicforecastsummary.htm

http://www.haber7.com/haber/20100607/Azerbaycanla-Sahdeniz-projesi-imzalandi.php

http://www.ekurd.net/mismas/articles/misc2012/5/invest841.htm

http://www.bp.com/sectiongenericarticle.do?categoryId=9006669&contentId=7015 093

http://www.abgs.gov.tr/files/AB_Iliskileri/AdaylikSureci/IlerlemeRaporlari/Turkiy e_Ilerleme_Rap_2003.pdf

Declaration – Prague Summit http://www.eu2009.cz/en/news-and-documents/press- releases/declaration---prague-summit--southern-corridor--may-8--2009–21533/ http://www.bilgesam.org/tr/index.php?option=com_content&view=article&id=2158:doalgaz-boru-hatlar-projelerinde-bueyuek-oyun-nabucco-gueney-akm-seep-ve-tanap&catid=131:enerji&Itemid=146

Gazprom: Blue Stream project between Turkey and Russia http://gazprom.com/production/projects/pipelines/blue-stream/

http://europa.eu/rapid/press-release_IP-12-721_en.htm?locale=en

http://www.euractiv.de/energie-und-klimaschutz/artikel/gaspipeline-deal-tuerkei- und-aserbaidschan-fr-tanap-006466

http://eurodialogue.org/Big-Game-heating-up-on-Caspian-gas-with-Turkey-Turkmen-move

http://www.jamestown.org/single/?no_cache=1&tx_ttnews[tt_news]=38834

http://www.upi.com/Business_News/Energy-Resources/2012/07/05/TANAP-pipeline-to-bring-gas-to-Bulgarian-border/UPI-83161341502061/

http://europa.eu/rapid/press-release_IP-12-1041_en.htm

http://www.europeanenergyreview.eu/site/pagina.php?email=faik_61%40hotmail.com&id_mailing=302&toegang=577bcc914f9e55d5e4e4f82f9f00e7d4&id=3455

Republic of Turkey Ministry of Energy and Natural Resources: Natural Gas http://www.en erji.gov.tr/index.php?dil=tr&sf=webpages&b=dogalgaz&bn=221&hn=&nm=384& id=40694,(accessed on 14/11/2011)

Republic of Turkey Ministry of Energy and Natural Resources : Natural Gas http://www.enerji.gov.tr/index.php?dil=tr&sf=webpages&b=dogalgaz&bn=2 21&h n=&nm=384&id=40694

Reports

Bundesministerium für Wirtschaft und Technologie/Bundesministerium für Umwelt, Naturschutz und Reaktorsicherheit: Energieversorgung für Deutschland. Statusbericht für den Energie gipfel, Berlin 2006.

Bundesministerium für Wirtschaft und Technologie: Energie in Deutschland, Trends und Hintergründe zur Energieversorgung, 2010

BP: Statistical Review of World Energy, June 2010 BP Statistical Review of World Energy, June 2011

DELOITTE Consulting (2008); Energy Report of Turkey

Directive 2003/87/EC of the European Parliament and of the Council,

Directive 2006/32/EC of the European Parliament and of the Council of 5 April 2006.

Deutsches Nationales Komitee, Weltenergierates Energie für Deutschland 2006

Council of European Union; Council conclusions on strengthening the external dimension of the EU energy policy, Press release, Brussels, 24 November 2011

EMRA (2012); Natural Gas Market 2011 Sector Report. European Commission, Turkey 2005 Progress Report, Brussels European Commission, Turkey 2007 Progress Report, Brussels European Commission, Turkey 2008 Progress Report, Brussels EU trends to 2030, European Commission 2009

EPDK: Turkish Energy Market: An Investor's Guide 2011 EPDK: 2010, Electricity Mark Report of Turkey

ETKB – Documents about Nuclear Reactor Planning of Turkey

European Commission, 2008i Joint Progress Report by the Council and the European Commission to the European Council on the Implementation of the EU Central Asia Strategy,

European Comission: Market Observatory for Energy 2011

Türkiye Enerji ve Enerji Verimliliği Çalışmaları Raporu "Yeşil Ekonomiye Geçiş" Temmuz – 2010

TÜBITAK; Vision 2023 Technology Foresight Project, Energy and Natural Resources Report

Turkey 2010–2011 Progress Report

The Energy Charter Treaty and Related Documents: 2004; A Legal Framework for International Energy Cooperation, Brussels, Energy Charter Secretariat.

Publication: Eurasia Daily Monitor, Volume: 9, Issue: 98

Proposal for a Directive of European Parliament and of the Council: 2002, concerning measures to safeguard security of natural gas supply, 2002/0220 (COD)

Premiership Liberalization Administration: Electricity Energy Sector Reform and Liberalization Strategy Document

Republic of Turkey Ministry of Energy and Natural Resources Annual Report 2009

Republic of Turkey Ministry of Energy and Natural Resources, Blue Book 2011

Republic of Turkey Ministry of Foreign Affairs: Turkey Energy Strategy, May 2011

Renewable Global Status Report

National Bureau of Asian Research (NBR): 2007, The Rise of Asia's National Oil Companies: Competitive and Geopolitical Implications. NBR Conference Report

IAEA Nuclear Technology Review 2010

International Energy Agency Geothermal Implementing Agreement, Strategic Plan: 2007

International Energy Agency: World Energy Outlook, 2010 International Energy Agency: 2008, World Energy Outlook

International Energy Agency: 2012, World Energy Outlook IEA Energy Review Turkey, 2009

International Energy Agency Geothermal Implementing Agreement, Strategic Plan: 2007

World Energy Council Turkish National Committee: Energy Report Ankara, 2010 World Energy Council, 2010 Survey of Energy Resources

National Bureau of Asian Research (NBR) (2007); The Rise of Asia's National Oil Companies: Competitive and Geopolitical Implications. NBR Conference Report.

Annex: Interviews

Within the scope of this dissertation, an interview was conducted on 10/09/2012 with Mr. Fazil Senel, the former general director of BOTAS, the public participation company of Turkey, and member of board of directors of EPDK, on the importance of the Southern Gas Corridor in Turkey–EU relationships and the impact of the TANAP project.

– *Could Turkey use its role, which has been increasing within the framework of European energy security, as a strategic card in its relationships with the EU?*

The European Union wants to diversify and create alternatives for its natural gas need, which has been increasing the last 20 years but has become stagnant in recent years, through different resources and thus to create a competitive structure. Its most important route to get rid of the Russian hegemony is through Turkey. Because, around 70% of the hydrocarbon resources of the world such as natural gas and oil, are located in the geographical area on the east and south east of Turkey. This naturally puts Turkey on the agenda as a non-alternative and strategic route.

– *Turkey has been carrying out negotiations with North Iraq regional administration on delivering the gas and oil resources to the EU. At this point, is it a near possibility that North Iraq resources could reach to the EU market? If so, under which projects could the resources be delivered to West?*

Turkey aimed towards zero problems with the neighboring countries, considering them commercial partners. Currently Turkey is connected to Azerbaijan and Iran through natural gas lines and Azerbaijan and Iraq through oil pipelines. A cabinet decree has been issued by the private sector for the construction of a natural gas pipeline that will reach from Iran to Europe. The "Arab pipeline" project, which would ensure natural gas connection with Syria, was completed, could not be commissioned due to the conflict in Syria. Therefore, projects such as delivering new oil and natural gas pipelines from the north of Iraq to global markets over Turkey are being negotiated so as to not disturb the integrity of our neighbour, and the finalisation of these negotiations is totally dependent on the conjecture and the political stability in our region. In my opinion, agreement is very soon.

– *Is the target of Turkey, to become the fourth biggest main gas artery of Europe after Russia, Nigeria and Algeria, realistically taking into account the existing energy policies of Turkey?*

Within the framework of the issues and the realities that I have indicated above, it is a totally realistic target to be reached with political stability and strong national will. In our opinion, approximately 120 bcm gas will be carried as transit over Turkey in 15 years.

– *According to the scenarios, the energy consumption will increase in the coming 20 years; it is estimated in particular that 80% of the EU will be dependent on gas from abroad. Within this framework, Turkey provides 45% of its electricity generation from gas. What are Turkey's plans to decrease dependencies for gas from abroad and what type of contribution is provided to the EU's energy security?*

Turkey is determined to reduce the share of natural gas in energy generation to 20%. In the near future, the aim is that natural gas, nuclear, oil, hydroelectric and renewable resources will have equal weight in energy generation. The principle target in the long term is an environmentally friendly energy generation where the renewable energy resources have a great share.

– *The BTC, BTE, ITGI and the TANAP–TAP projects strengthened the position of Turkey as a regional energy hub. Will such projects accelerate Turkey's EU accession process and do the projects have any impact on Turkey being a candidate country?*

Such projects are leading projects, which will be followed by bigger, more important ones. It is of no doubt that they are the most important alternative projects of Europe, therefore their contribution in the candidacy process is in a positive direction.

– *The European Commission supports the opening of the Southern Gas Corridor. Shah Deniz Consortium selected Nabucco West at the first stage, and determined TAP as the exit route in the final elimination. After this stage, what do you think will be the end of Nabucco West?*

We are of the opinion that the project named Nabucco West will be realised at future stages although not now.

– *The cost in the construction of the South Stream developed by Gazprom exceeded estimations. Is the increase of costs a hindrance for the realisation of the project? Most of the Eastern European countries are highly dependent on Russian gas. What sort of project could be developed in the diversification of energy supply towards that region?*

Since the cost of the South Stream project will be paid by the customers through long-term rates, I think it's appropriate to remind ourselves that the realisation of this project is totally dependent upon the demand of the customers.

An interview was carried out with the economy attaché of Azerbaijan to Berlin, Germany, Elman Muradov, on the Southern Gas Corridor and TANAP project.

– In recent years, Turkey and Azerbaijan have been developing many joint projects including initially the Baku–Tbilisi–Ceyhan oil pipeline and then Shah Deniz I and II projects. In this regard, to what extent could Turkey and Azerbaijan contribute to the EU's energy security in strategic terms?

Turkey is in the position of being a corridor country for the transmission of energy resources (oil and natural gas) from the Caspian Basin and Middle East countries, such as Azerbaijan, Russia, Iraq, Iran, Egypt and Turkmenistan, to the international market. Besides following the extraction of gas on the West Nile Delta by BP in January 2007, it could play a significant role in the transportation of Egyptian gas. In addition to these, Turkey could become a key country in transporting liquefied gas through vessels to international markets.

The geopolitical position of Turkey, which is highlighted in the context of energy security, is at a level that can enable beneficial collaboration between Ankara and the EU. The most important aspect of this process seems to be related to the ability of Ankara to engage in new initiatives that put the emphasis on the determinative role of energy or, to orientate the existing initiatives in a controlled way. When considered in particular within the context of Turkey–EU relationships, Turkey is required to turn the advantages granted by being a safe route country in the context of the benefits afforded to EU countries from Middle East and Caspian Basin energy resources, into action through the pipelines it will construct within a short period.

– Taking into account the fact that the future supply-demand balance between the EU and Turkey and Azerbaijan and the EU will increase, to what extent will interdependency affect diplomatic relationships? Within this framework, do you think that the opening of the Southern Gas Corridor will accelerate Turkey's EU membership period?

Energy has a unique position in the establishment of the EU. The developments experienced in the coal market that dragged the countries into war created a reference point for the establishment of such a union. Following the current expansions, the EU continued to attach importance to its relationships concerning energy. At this point, EU–Turkey relationships took on a significant character since Turkey is a corridor that reaches both to the Middle East and Central Asia. In particular, there exists analysis suggesting that from 2020 onwards the EU, which originated from energy, will become more dependent upon energy than it is now. This dependence refers to the risk the EU faces in losing Turkey, a strategic partner. It is obligatory for Turkey to attach great importance to its relationships with Georgia, Azerbaijan, Iraq and Central Asian countries. It should present various diplomatic activities that could widen its bilateral relations. Making the EU-Turkey relationship interdependent will make the European identity stronger in the foreign policy.

– *Whilst Shah Deniz Consortium has selected Nabucco West at the first stage, it determined TAP as the exit route in the final elimination. What do you think will be the end of Nabucco West after this stage? Also, what are the reasons, according to you, for the preference of TAP?*

One of the reasons that were effective in taking this decision were the active policies being employed by those on the TAP side. In addition to the active policies carried out by the parties in Shah Deniz Consortium, it was claimed that bureaucratic problems were experienced in West Nabucco and that there was no one single voice among the partners. As a matter of fact, until the December of 2012, the possibility of West Nabucco winning this contest was quite high. Thereafter, more initiatives and actions taken by TAP led to this result.

Energy has both political and economic aspects. However, it is appropriate to note one point. TAP is a project that satisfies all stakeholders having an interest in the region, in particular Azerbaijan. This is such that although this line is an alternative project to Russia, this also has an important impact on the interests of its northern neighbour.

Azerbaijan is also content with this agreement. That is, TAP is 450 km shorter than Nabucco and thus is economically more effective. The construction costs of the arch will be significantly less compared to Nabucco West. Azerbaijan will now market its own natural gas directly to the European market over global market prices. This means that a new wide market will be opened for Azerbaijan, that market dependence on Russia will end and that Baku will have reached its basic purpose in natural gas exports. Whilst Baku makes significant contributions in energy security of the European Union with this decision, it refrains from competing with Moscow, thus protecting its good relationship. From this aspect, TAP agreement has both political and diplomatic importance for Azerbaijan.

– *Is the construction of the South Stream pipeline in danger? Has the construction of this pipeline changed the position of Turkey in the energy resources?*

Following the permission granted to the construction of South Stream, the Russian-Turkish relationship in the field of natural gas transit has acquired a new character. Since Turkey is available as a center of distribution for energy resources due to its geographical position, it could be used to support the Nabucco Project at the initial stages; however, it does not support this project due to political and national reasons. Due to unsolved problems, Nabucco has existed only on paper for some years and the possibility of putting this project into life has been decreasing over time. Although Turkey agreed with the construction of South Stream costing around 30–35 billion USD, the realisation of the project seems to be hard in economic terms. Rather than the completion of this project, it would be more appropriate for Turkey to increase the capacity of Blue Stream, which would also benefit both countries. As a matter of fact, thanks to the pipelines system of Turkey, Russian natural gas could be delivered to the European market at a cheaper price compared to South Stream. Russia could strengthen its position as a supplier by spending less and Turkey would come a step closer to becoming the energy corridor of Europe.

An interview was held with Mr. Şahin Arıkan, who was assigned as technical head of department in Nabucco International Company in Vienna, on Turkey's energy politics and Nabucco.

– *You worked as Deputy General Director of BOTAS, and since 2009 you have been working as a technical head of department in Nabucco International Company. How would you define the role of energy in the Turkey–EU relationship and its impact on the membership?*

One of the three fundamental reasons used by those who support Turkey's accession to the EU is the contribution to the EU's energy security through the energy corridors to be created via Turkey. The two others are the contribution to the alliance of civilisations and peace, as well as the young labour force. Energy security has a vital importance for the EU, which is one of the biggest energy importers of the world, and in particular with the withdrawal of Germany from nuclear energy, the importance of natural gas has increased more with the decreasing production.

On a similar note, breaking the dominant position of Russia, the largest importer for the EU, could only be possible by creating continuous and secure alternative corridors. This being the case, the routes reaching the Middle East and the Caspian area, which have the biggest gas and oil reserves of the world, pass through Turkey. This in turn strengthens the geopolitical position of Turkey. Considering the fact that the issue of energy is among the top agendas within the context of international relations, it is quite natural that it has a determinative role in Turkey–EU relations.

– *Can you elaborate on the policies developed by Turkish foreign policy in the field of energy?*

This issue has been one of the foreign policy targets of Turkey in the last quarter century and various projects have been developed towards this target. The first of these is the Baku–Tbilisi–Ceyhan Oil Pipeline, which came to the top of the agenda after the separation of Azerbaijan from the Soviet Union. The international agreements of this project, negotiations for which were launched in 1992, were finalised in 1999 and the project was completed in 2006 after a determined and successful attempt and so the first supply of oil was transferred to the Mediterranean. Ten thousand and seventy-six kilometers of this pipeline with the capacity of one million barrels per day pass through Turkish territory and are being successfully operated. A significant part of the oil is loaded to tankers from Ceyhan and is transported to European ports. Turkey has continued its efforts to remain as the main artery of natural gas after oil, and in this context 210 km of the Turkey–Greece Natural Gas Interconnection Project, which is one of the projects that the EU has developed within the scope of INOGATE Programme, passes through Turkey and 85 km thereof through Greece which resulted in gas supply to Greece in 2007. The third stage of this target is delivering the gas beyond Greece and/or Bulgaria, which has newly been put into life as the Trans Adriatic Project and was selected as the export route by Shah Deniz Consortium as part of the South European Natural Gas Interconnection project. Briefly, what is spoken is when the projects that are ongoing will be completed rather than whether these policies are realistic.

– *You have been working in Nabucco International Company in Vienna since 2009. Could you please elaborate on the past and future of the Nabucco project?*

Shahdenizi partners indicate that the decision is not a strategic one, but is totally technical and economic. They state that it is a shorter line, with a cheaper transportation cost and also that they received a more appropriate offer than Nabucco from the gas purchaser at the end of the pipe, adding that this is a main impact in the selection of TAP. Nabucco was a project which could have been a balancing element for the market in Central and Eastern Europe. Unfortunately, Europe seems to have lost this opportunity. It seems that the dominant position of Russia in the market will continue. Nabucco partners indicate they will not continue with the project without any commitment for gas procurement and they have even put the decisions in this direction into practice. Nabucco West or Nabucco Classic, in short Nabucco will suspend its physical activities and continue as a "shelf company" for some time, until the old or new partners take *this* project off the shelf. If we do not need Nabucco in the short and middle term, Nabucco will be history.

– *Does Nabucco West have any chance to be implemented in the future, or could other projects be developed?*

As it is indicated above, the partners have taken the decision to suspend company activities. If, in the future, EU and/or gas importing countries/companies develop important projects in order to transport Caspian and Middle East gas to the market Nabucco addresses, Nabucco or a new project on the Nabucco route could come into life.

– *What is the level of EU-Turkey relationships? To what extent does the energy and interdependency play a role in these relations? Does any other factor play a role?*

The negotiations between the EU and Turkey have long been suspended; even the energy chapter was not opened despite direct and indirect requests from Turkey. I think the reason for this could be the cutting off of the Turkish part of the Nabucco Project and it being organised under a new body named TANAP. It is of no doubt that these pipeline projects passing over Turkey have and will have great contribution in increasing and strengthening the security of the EU. The parties will continue with their relationship with the principle of mutual interest. The negotiations will not be interrupted in the projects developed in this direction and the European countries will continue to develop their activities in Turkey from various aspects. Today, billion dollar investments are being made in Turkey by many companies such as E.On, GDF, RWE, OMV, Shell, BP and in line with the stable administration and regulations that encourage investments, these relationships will gain strength with the principle of interdependency and such companies will increase their presence in Turkey, which they consider as a strategic partner for their Middle East and Central Asia operations. These relationships and the interest paid will have positive impacts on the mutual relations between the EU and Turkey.

www.ingramcontent.com/pod-product-compliance
Ingram Content Group UK Ltd.
Pitfield, Milton Keynes, MK11 3LW, UK
UKHW021324180426
11947UKWH00017B/1426

9 783631 744789